Cosmic Horizons

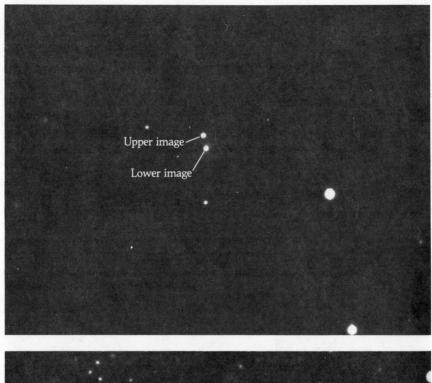

Upper image

Lower image

Upper image

Lower image

This is the first known example of the gravitational lens effect, which was predicted in 1936 by Einstein on the basis of his general theory of relativity. For a more complete description, see page 190.

Cosmic Horizons

Understanding the Universe

Robert V. Wagoner
and Donald W. Goldsmith

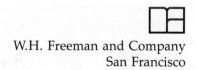

W.H. Freeman and Company
San Francisco

This book was published originally as a part of *The Portable Stanford,* a series of books published by the Stanford Alumni Association, Stanford, California.

Library of Congress Cataloging in Publication Data

Wagoner, Robert V.
  Cosmic Horizons.

  Reprint. Originally published: Stanford, Calif. :
Stanford Alumni Association, 1982. (The Portable
Stanford)
  Bibliography: p.
  Includes index.
  1. Cosmology. 2. Astronomy. I. Goldsmith,
Donald. II. Title. III. Series: Portable Stanford.
QB981.W233 1983   523.1   82-20994
ISBN 0-7167-1417-5
ISBN 0-7167-1418-3 (pbk.)

Printed in the United States of America

84-4879

9 8 7 6 5 4 3 2 1 MP 1 0 8 9 8 7 6 5 4 3

To Lynne, Alexa, and Shannon
and to Rachel
for their patience, love, and understanding

CONTENTS

PREFACE

WITH THIS BOOK we seek to involve a wider audience in a great and continuing adventure: the exploration of the most remote frontiers, the vast reaches of space and time that encompass the known universe. We therefore emphasize cosmology, the study of the universe as a whole, rather than the many other aspects of astronomy—the particular properties of the constituents of the universe (such as the solar system, stars, or galaxies). We have been guided in our endeavor by the fundamental question: To what extent can we understand the universe? We must investigate the horizons that limit our ability to probe nature, from its particle constituents to the evolution of all structures revealed to us as we extend our view out into space and therefore back into time. How have we expanded these horizons in the past, and which can we hope to expand in the future?

Cosmologists and astronomers are not only explorers of new frontiers, they are also in a sense detectives. Their tools fill laboratories and observatories around the world and in orbit above our planet. The mental framework for their deductions is constructed from the laws of physics. Their clues are the various forms of radiation that reach us from the cosmos and the nature of the matter around us. From these clues they seek to piece together the history of the universe and to sort out the prospects for its future.

This book emerged from a one-quarter course taught at Stanford University by R.V.W. The course aims to help undergraduates develop an understanding of the methods of science as applied to the subject of cosmology. This understanding, which we likewise seek to develop with this book, can come only from studying the subject in enough depth to reveal fully the logic of its development. To achieve this understanding, you must be willing to think through chains of closely reasoned arguments. It is our hope that your efforts as a reader will be rewarded with enjoyment as well as insight, as you expand your perception of our cosmic environment and history.

This book and our collaboration would not have existed without the support, encouragement, and skill of our editor at The Portable Stanford, Cynthia Fry Gunn. When our spirits flagged, her continuing dedication to excellence provided new motivation. We would also like to thank Elizabeth Campbell for her ongoing care in assisting with the production of this book, Rose Aleman for her careful typing of the original manuscript, and Karen Springen for her retyping of its many revisions. R.V.W. would like to express his appreciation to the Guggenheim Foundation for the award of a fellowship during the period when this project began. We are grateful to Halton Arp, Bernard Burke, Eric Feigelson, Owen Gingerich, Edward Harrison, Tobias Owen, and Jacob Zeitlin for their assistance in various aspects of this work.

Robert V. Wagoner
Donald W. Goldsmith

Stanford, California
Berkeley, California
August 25, 1982

Cosmic Horizons

1

Expanding Our Vision

THE PANORAMA OF A CLEAR NIGHT SKY emblazoned with stars has probably evoked the same questions in the minds of men and women throughout the ages. What is up there? Where have we come from and what is our fate? What role do we play in the universe? The search for meaning in our existence is linked to the quest to understand the nature of the cosmos.

The Quest to Know

These concerns have produced the science of cosmology, the study of the universe as a whole. Cosmology derives from the Greek words *kosmos*, meaning the world as an ordered whole, and *logos*, meaning discourse. Cosmology is thus the all-encompassing science, for it deals with the structure and evolution of the entire universe—everything that we now observe and that we can ever hope to observe in the future. Astronomy, by contrast, deals with the properties of individual objects, such as stars and galaxies. Cosmology provides the canvas on which the detailed properties of our world are painted by our knowledge of physics, chemistry, biology, and the other sciences. Not surprisingly, cosmology is one of the oldest of human intellectual endeavors; it responds to the same questions that led to the development of religion and philosophy.

Although mankind has long been concerned with the questions of cosmology, our scientific study of the universe is the inheritance of a tradition of ancient Greek thought, refined by many great intellects during the past two millennia. Only during the past two generations, however, have astronomers and physicists begun to feel that a comprehensive picture of the universe is beginning to emerge. As we shall see, this confidence rests on a key assumption, the principle that our view of the universe is a *representative* one. With this principle, we can discuss the properties of the

1

universe beyond the borders of our vision. Without it, we must remain ignorant of the universe as a whole.

Astronomical observations during this century have led us to a radically new view of the cosmos. The classical picture of a timeless, unchanging universe has been replaced by the view of a universe evolving on all scales, from the fundamental constituents of matter to the distribution of galaxies. As we look out in space, we are looking back in time, for we see any source of light (or any other form of radiation) as it was when it emitted the radiation we receive today. We have discovered that the most remote regions of the universe appear different from those nearby—not because the properties of the universe change with distance, but because they change with time. Research in cosmology has entered a new era, one in which we are probing the past, observing directly the evolution of the universe. This critical new aspect of cosmology—the expansion of our horizons back in time as well as out in space—will provide the focus of this book.

Why Is Cosmology Unique?

Because of the breadth of its scope, cosmology has several unique aspects that make it fundamentally different from the rest of science. This uniqueness forms a deep and subtle property of cosmology.

Along with most of astronomy, cosmology deals with things forever remote from our direct experience. We have visited, either by manned or automated spacecraft, our closest celestial neighbors within the solar system, and future generations might reach other stars (though the distance to the closest stars beyond the sun exceeds the distance to the moon by 100 million times). Despite the great distances we have traveled, the region of space accessible to us forms just a tiny bubble in the vast cosmic sea.

Therefore, unlike all other branches of science, cosmology (and most parts of astronomy) does not yield directly to experiment. We can never bring the object of study (the universe as a whole, let alone a star or a galaxy) into our laboratory. The universe is not under our control. We must rely on information astronomers receive in the form of light and other radiation. Our knowledge is indirect, decoded from the patterns of radiation detected by observatories all over the earth and in orbit around the earth. Because of this remoteness, interpretation of data of necessity plays a more critical role in cosmology than in most other fields of science. Observations of remote objects provide much less information than does direct experience. If you can take an object into your laboratory and analyze it with a variety of specialized instruments, you can learn far more about it than you can at an immense distance.

A second aspect of cosmology that makes it unique is that we are part of the system being studied. We cannot remove ourselves from the universe and view it as detached observers. To what extent this circumstance affects our ability to understand the universe fully remains unclear.

A third way in which cosmology is unique has been touched on: We see objects as they were, not as they are. The radiation that brings us news of the universe travels through space at the speed of light, 300,000 kilometers per second. Great as this speed is, light and all other forms of radiation still take many years to travel between the stars. When we look outside our family of stars, the Milky Way galaxy, we look at parts of the universe so distant that information from them takes millions to billions of years to reach us. We can never hope to know directly what these remote regions of the cosmos are like today or were like even thousands of years ago. Instead, we must deal with out-of-date information.

But in return for being out-of-date, the radiation from great distances brings us news of the past. The farther we look out into space, the further back we see into the past. We can think of objects that we observe at greater and greater distances as embedded within imaginary spherical shells of increasing radius, centered on us (see Figure 1-1). Each more distant shell contains objects as they existed at correspondingly earlier times. As we look outward from the earth, we may think of ourselves as located at the center of a giant onion, which we peel from the center outward as the limit of our telescopes is extended to greater distances. Every distance from us represents a corresponding time prior to the present moment—the time that it has taken radiation to travel that distance to us.

Finally, cosmology is unique in that it provides only one object of study. Astronomers can compare stars, physicists can compare different types of particles, biologists can compare cells. But cosmologists, by definition, have available only one example of their subject. We cannot compare different universes in order to discover which features are general and which features are specific to our universe. Thus we can never know for sure whether our universe has a particular property, such as the existence of life, by chance or by necessity. Nevertheless, there have been attempts to answer the question, "Why is the universe the way it is?" We, too, shall address this question in the final chapter.

The uniqueness of cosmology makes it an intriguing subject but produces many pitfalls, which we shall encounter again and again on the road to our understanding. Can our finite minds fully grasp the immensity of the universe, both in space and in time? We have difficulty constructing a mental image that encompasses the universe. But this is precisely what we must do if we are to grasp the basic elements of cosmology.

How Do We Understand the Universe?

The process of obtaining knowledge of the universe involves the interplay of three activities: (1) laboratory experimentation, (2) mathematical

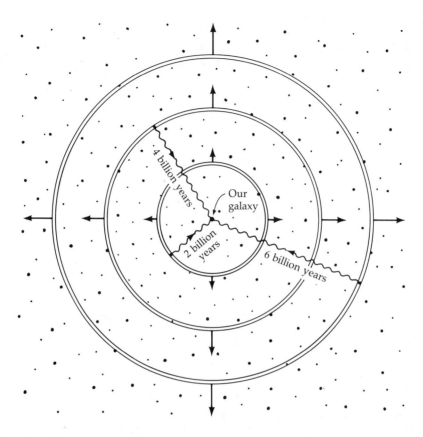

Figure 1-1. We can imagine objects in the universe (such as galaxies) as contained within spherical shells, each at a certain distance from us. The light (indicated by wavy lines) reaching us from a given distance brings us information from a corresponding time in the past, namely the time for light to travel that distance. (Later, we shall learn that these shells are expanding.)

formulation and application of the laws of nature, and (3) observation of distant objects (see Figure 1-2). The exploration of the universe actually begins in the laboratory. For instance, in order to understand the nature of the universe we must understand the nature of elementary particles, which are often determined by accelerator experiments. In the laboratory, scientists struggle to discover the rules that govern the nature and behavior of all forms of matter—the laws of physics. Without these laws we could not begin to make sense of the information we obtain from the cosmos. With telescopes, we can map the distribution of radiation at various frequencies over the sky. The laws of physics enable us to convert these raw data into understanding: They provide a model of how the universe works.

For example, although the light we receive from stars comes only from their surfaces, our analysis of the characteristics of that light allows us to develop a picture of stellar interiors and an understanding of how stars evolve in time.

How do we discover these laws, which we believe govern all aspects of the physical world? The process that has been proven best involves the intimate interplay of experiment and theory. A scientist (or scientists) proposes, in the form of a mathematical theory, a law governing a particular aspect of nature. The consequences of a variety of such proposed theories are then derived as predictions for the outcome of experiments. Those theories whose predictions fail to agree with even a single experiment must be discarded or modified. Among those theories then remaining, we usually choose the theory that is mathematically simplest, because experience has taught us that nature seems to operate that way. The laws of nature are simple when expressed in the proper mathematical form—a profound fact.

Gravitation affords an example of the process by which we develop laws that are increasingly accurate representations of nature. In 1916 Albert Einstein proposed his theory of general relativity as a more accurate and comprehensive description of gravitation than the prevailing Newtonian theory. One of Einstein's predictions was that light (or any other form of radiation) would be bent by a gravitational field, whereas Newton's theory did not even address this question. Einstein predicted the exact amount by which the apparent position of a star would be shifted by this bending when the starlight's path passed near the sun (see Figure 1-3).

Expeditions were soon organized to test this startling new prediction during a total solar eclipse, the only time when stars close to the direction

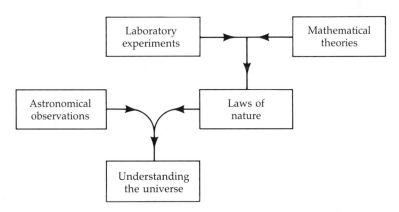

Figure 1-2. This block diagram shows the scientific process of obtaining knowledge of the universe.

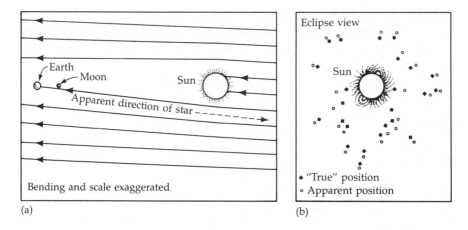

Figure 1-3. (a) The rays of light from a single star are bent by the sun's gravitational field. The closer a ray passes by the sun, the more it is bent. (The position of the moon during an eclipse of the sun is also shown.) (b) When we look toward the sun at a time of total eclipse by the moon, when stars near the sun are visible, this gravitational bending shifts the stars' positions on the sky from their "true" positions. We can measure the true positions some months later, when the sun no longer lies in that direction on the sky.

of the sun can be seen. When the results of these observations were found to agree with Einstein's prediction, they not only created a sensation but, more importantly, they overthrew our previous Newtonian conception of the nature of gravitation. Modern, more accurate observations still agree with Einstein's predictions and have disproven most other competing theories. Yet we have no reason to believe that Einstein's theory will not be replaced eventually by another. The testing continues.

When we apply the laws of nature to our observations and to the construction of models of astronomical phenomena, we make a critical assumption. We suppose that the laws that control matter in our laboratories have equal validity in the most remote epochs and the most distant regions of the universe, where physical conditions can be far different from those we encounter within the solar system today. Fortunately, we have uncovered no clear evidence that contradicts this assumption. For example, the patterns of spectral lines (the excess or deficiency of light at certain wavelengths) in the light from the atoms in the most distant objects we have seen, the quasars, are identical to those produced by the same atoms on earth. This suggests that the laws governing the structure of atoms are universal.

How do astronomers gather information about the universe through their observations—the third element in the process of understanding? Until very recently, our view of the universe was restricted to that provided by visible light, which forms only a small portion of the spectrum of electromagnetic radiation (see Figure 1-4). But during the last 30 years astronomers have begun to "see" the universe in the "light" of virtually all other wavelengths of electromagnetic radiation: gamma rays, X rays, ultraviolet radiation (UV), infrared radiation, microwaves, and radio waves. This revolution in our ability to probe the universe in different ways has been made possible by the development of new detectors and observational techniques, such as radio telescopes, X-ray telescopes, and cooled infrared detectors. Also important has been the ability to place telescopes in orbit around the earth, where radiation of those wavelengths that cannot pass through our atmosphere (see Figure 1-4) can be detected.

The recent history of astronomy has shown that as each of these windows to a different form of electromagnetic radiation has been opened, new and unexpected phenomena have been discovered. Some examples are quasars, pulsars, X-ray sources, and the sea of photons believed to be

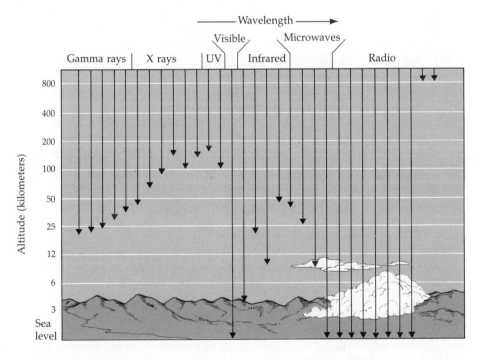

Figure 1-4. Electromagnetic radiation exists over a wide range of wavelengths, called the electromagnetic spectrum. Our atmosphere absorbs radiation of different wavelengths to different extents. Of the radiation emitted by cosmic sources, only that portion within certain regions of the spectrum (primarily visible and radio) can reach the ground. Arrowheads indicate the altitude at which radiation of each wavelength becomes completely absorbed by the atmosphere. Note that wavelength increases to the right.

a remnant of the early universe. The radio sky turns out to look quite different from the sky as seen in visible light; the infrared sky bears little resemblance to either; and the X-ray sky is stranger still. We shall in later chapters consider each part of the electromagnetic spectrum in order to appreciate what astronomers have discovered about the universe.

Yet these forms of electromagnetic radiation are by no means the only signals that we may receive from the universe. As we shall see, other types of radiation are potentially detectable. One is gravitational radiation, produced by the motion of masses in a manner similar to that by which electromagnetic radiation arises from the motion of electric charges. The promise of this new probe is great, for it has the potential to provide us with information about regions of the universe that can be viewed in no other way.

Observing the radiation that comes to us from very distant objects is not, however, the only way we can learn about the universe. Another method exists, which involves what seems to be a simple activity: examining the matter in our vicinity of the universe. We might describe this study as "cosmic archaeology." We have learned that the constituents of the present universe contain clues to its past history. The forms in which matter appears today reflect its processing through the variety of physical conditions to which it was exposed in the past. For instance, an accurate determination of the abundances of certain nuclei today may tell us something about conditions in the universe billions of years ago.

All of these probes, the various forms of radiation and relics, allow us to expand the horizons of our knowledge of the universe. In addition, our increasing knowledge of the laws of physics, coupled with our increasing ability to apply these laws to astronomical objects, permits us to extract an ever-more-profound understanding of the universe from these probes.

The Cosmic Uncertainty Principle

In the universe, as on earth, proximity leads to familiarity. Of necessity, our knowledge of remote objects is less complete than our knowledge of objects nearby. This problem becomes particularly severe in the vast, uncharted regions of the universe. The farther out we look, and thus the further back in time, the greater is our uncertainty concerning the true nature of the universe.

The uncertainty associated with any experimental or observational result is related to the accuracy with which some quantity can be determined. For instance, consider the measurement of distance. Distances within the solar system have been determined to an accuracy of far better than 1 percent, whereas the distances to even the nearest stars are uncertain by at least a few percent. For more distant stars within our galaxy, the uncertainty in distance increases to more than 10 percent. The distances to nearby galaxies are also uncertain by more than 10 percent, and the uncertainty increases still further for more distant galaxies.

Greater distance leads to greater uncertainty in our knowledge of the universe for a number of reasons. The apparent brightness of an object, for example, decreases with greater distance. Because the brightness of a source determines the amount of information we receive, less brightness leads to less completeness and accuracy in our analysis. In addition, the greater the distance, the greater the chance that intervening material such as dust and gas will modify the radiation during its passage from the source to us, confusing or obliterating its message.

As we explore the universe to ever-greater distances and more remote times, it is important to keep this uncertainty principle in mind. Our knowledge of the farther reaches of the cosmos will always remain less secure than that of our more immediate cosmic neighborhood.

Given this cosmic uncertainty principle, how confident can we be that any picture of the universe as a whole is correct? Can we hope to understand the nature of objects so distant that the light by which they appear to us today was emitted from those objects billions of years ago? Can we hope to understand the nature of the universe at remote times in the past when no life, no planets, no stars, and no galaxies existed? Can we hope to be able to predict its future evolution? Can organized structures of molecules on a run-of-the-mill planet orbiting a typical star in an ordinary galaxy really comprehend their cosmic environment?

In this book we shall attempt to describe how far we have progressed toward answering such questions. If we are successful, you will not only see and appreciate how much has been learned about the universe but also sense the challenge of how much remains to be discovered. The contrast between the vastness and complexity of the universe and the finite limits of our minds assures us that many intriguing questions will always await an answer, with many more questions waiting to be asked. For as J.B.S. Haldane said in 1927, "The universe is not only queerer than we suppose, but queerer than we *can* suppose."

2

Changing Concepts of the Cosmos

FOR COUNTLESS GENERATIONS, our ancestors have marveled at the mysteries of the night sky. Some of their awe and imagination has come down to us; some has been lost as we have focused our attention on the civilizations we have created. As our scientific understanding has increased, our conception of the nature of the cosmos has changed greatly. The rate of conceptual change has accelerated most dramatically during the past century.

We may characterize the evolution of our picture of the cosmos as progressing from a *flat,* to a *round,* and finally to a *uniform* model of the universe. This progress reflects our growing awareness of the vastness of the universe and the relative insignificance of our immediate surroundings. Since our forebears in ancient times possessed no evidence to contradict the apparent "fact" that the earth was flat, they assumed that celestial objects existed "above" us. The round view placed the earth (and later the sun) at the center of a universe in which celestial objects moved in circular orbits. Finally, the uniform model recognized, first, that our sun is but one of many stars that comprise the Milky Way, and, eventually, that the Milky Way is but one galaxy in a sea of galaxies that fill the universe. In what follows, we shall see how these views developed.

Origins of Cosmological Thought

We have only fragmentary knowledge of how our ancestors envisaged their relation to the universe before the Greek civilization that dates from 600 B.C. However, we do have some evidence of the various methods ancient cultures used to observe the heavens. For thousands of years our forebears studied the motions of celestial objects and tried to perceive the patterns of these motions.

The oldest astronomical records we possess are prehistoric bones carved with markings that apparently represent the phases of the moon. Some of

Figure 2-1. At Stonehenge, in England, great slabs of stone form a circle inside the original astronomical markers, no longer present. On midsummer's day (June 21 or 22) an observer at the circle's center would see the sun rise over the "Heel stone," visible beyond the circle. Other alignments record the rising and setting of the moon at the extreme northerly and southerly points of its motion. (Photo: Owen Gingerich.)

these bones have ages of 30,000 years and resemble the calendar sticks made by aborigines of our modern era. The earliest coherent evidence we have of astronomical observations consists of tablets from the Mesopotamian civilization that flourished during the two millennia after about 4000 B.C. These tablets, which record the motions of the moon, sun, and planets against the background of stars, represent observations that reach the limit of accuracy attainable without the aid of instruments. The Chinese began keeping written notes of astronomical events at about the same epoch, followed in time by the Hindus, Egyptians, and Northern Europeans.

In addition to keeping written astronomical records, the ancients built structures to mark the movement of celestial objects. The great ring of stones at Stonehenge, England (see Figure 2-1) was built in stages during the late Neolithic Period and early Bronze Age—between about 2000 and 1400 B.C. This impressive monument served as some sort of temple for sky worship and clearly functioned as an astronomical observatory and record keeper, as becomes apparent every summer solstice, when the sun on this longest day of the year rises directly above the "Heel stone," as seen from the center of the ring. Stonehenge is just one of many such megalithic observatories in the British Isles and in France. Other cultures built monuments for similar purposes. We do not know, however, what the builders of these ancient observatories thought about the cosmos. Did they look beyond the cycles of the heavens in search of rules to describe the physical reality they saw? Or were they content to trace the motions of objects that seemed totally different from earth—remote and mysterious?

The latter view appears more likely when we consider what we have learned about primitive tribes that exist today. The !Kung Bushmen of the Kalahari Desert, for example, regard the heavens as a vast bowl suspended above the earth and the stars as campfires in the sky, the "backbone of the night." A similar view of celestial objects as different in nature from earth appears in ancient Greek philosophy as it developed during the first millennium B.C. Even as the Greeks applied their reasoning to the heavens to find the first "laws" of nature, they retained the belief that the sun, moon, stars, and planets differ fundamentally from the earth.

Greek Astronomy

Around the year 600 B.C., a remarkable development in the history of thought occurred in Ionia, the sprinkling of islands in the Aegean Sea and the neighboring mainland of what we now call Greece and Turkey. This change—a relatively rapid transformation in human attitudes about the world—marks the beginning of science. For the first time, so far as we know, humans came to regard the cosmos not as the home and plaything of mysterious gods, but rather as a universe accessible to reason, to the power of observation and deduction. We have little understanding of what caused this change in attitudes, but clearly the times were right for some people to speculate about the earth and the heavens in a coherent, orderly manner. From such speculation came startling conclusions.

Early Greek Thinkers

The earliest Greek astronomical record we have consists of a prediction by Thales of Miletus that an eclipse would occur in the year 585 B.C. and observations confirming this prediction, both described in Herodotus' famous history. Predicting such an occurrence required a long series of accurate records of past eclipses, and the mental ability to see a pattern within these past events. Collecting and interpreting such data would not in itself represent a stunning new advance in the ability to understand the heavens; the builders of Stonehenge may well have possessed similar ability. One can predict eclipses without understanding the nature of the sun and the moon. But Thales went on to ask: What comprises the universe? From this sort of wondering came a host of answers, many of them wrong. Thales postulated, for example, that the earth is a flat disk floating on water and that earthquakes result from waves in the water beneath the earth. But more important, from such wondering also came a host of new questions.

During the fifth century B.C., Anaxagoras, who moved from Ionia to Athens and helped establish Athens as the center of Greek culture, correctly concluded that the moon shines not by its own light but by the light it reflects from the sun, and that eclipses of the moon occur when the earth prevents sunlight from reaching the moon. We know next to nothing about how Anaxagoras reached such conclusions, for only a few fragments of his work survive. But the search for what we would now call a physical explanation distinguishes Anaxagoras' approach to the cosmos from the main trend of Greek philosophy, which led from Pythagoras to Plato.

Pythagoras, who lived between Thales' and Anaxagoras' lifetimes, was a great mathematician and mystic, as well as the founder of a short-lived religion, the Pythagorean brotherhood, which believed in reincarnation, vegetarianism, and the kinship of all living beings. The Pythagoreans perceived the visible world as striving for the perfection of an unseen mystical world, whose existence is revealed to us chiefly in numbers and in what we now call geometry. Enamored with the circle and sphere as the most perfect forms, the Pythagoreans were apparently the first to suggest that the earth is not flat but round. (Anaxagoras may have later reached the same conclusion by observing the shape of the shadow cast by the earth on the moon during a lunar eclipse.)

A century after Pythagoras' death (circa 497 B.C.), Plato developed a philosophy in which the real world consists of spiritual entities—"ideas," as Plato called them, or what we might call "ideal forms." These true entities can be understood by reason, not by observation. "We shall dispense with the starry heavens," said Plato in his *Republic*, "if we propose to obtain a real knowledge of astronomy." Thus the Pythagoreans and Plato diverted attention from observation while emphasizing the importance and power of logical reasoning.

In sharp contrast to Pythagoras and Plato, the philosopher Democritus proposed (around 400 B.C.) that the universe is simply matter and the void, and that matter consists of tiny, indivisible particles called "atoms." This "materialist" philosophy regarded atoms as eternally in motion, forming and reforming an infinite number of worlds. Democritus thus deserves credit as the first to suggest an infinite universe. He and his followers rejected the idea that the universe could be finite in extent, for what could form the boundary of the universe? And what would lie beyond the boundary? Democritus' ideas received scant attention from his successors in Greek philosophy, although we now know that his philosophy was much closer to the truth than that of most of his contemporaries and successors.

Aristotle's Cosmology

Plato's great successor Aristotle, who lived during the fourth century B.C., had more respect for what we call the "real" world than did Plato, believing that the visible world develops as "ideas" work on unformed matter. Careful observation of the visible can therefore lead to knowledge of the underlying reality. Adopting the Pythagorean tradition of finding perfection in spheres, Aristotle envisioned a cosmos of spherical shells centered on the earth. Four elements comprise our world—earth, water, air, and fire. Each of these four elements, said Aristotle, tends to assume its natural location, with the result that the heavier elements, water and earth, move downward. To Aristotle celestial objects consisted not of fire but of a fifth element, the "quintessence" or ether. Just where the air gives way to the ether in Aristotle's view remains unclear, but the transition from the quintessence to the familiar forms of matter seems to occur somewhere between the earth and the moon. The invisible spherical shells that hold the heavenly bodies rotate about the earth. Because Aristotle and his followers saw no motion of the stars relative to one another, they suspected (correctly) that the distances to the stars must be immense.

Aristotle imagined the universe not only to possess spherical symmetry but to be finite in extent. If the universe is not finite, he argued, one of its elements must be infinite, leaving no room for the others. However, we recognize today that an infinite volume can contain infinite amounts of different constituents.

Measuring the Heavens

Heraclides of Pontus, Aristotle's contemporary, was apparently the first to assert that the earth rotates, thus explaining why the stars and other celestial objects appear to move across the sky. A rotating earth, however, seems less likely to be the center of the cosmos than a fixed earth.

Almost a century after Heraclides, Aristarchus of Samos took this idea further by suggesting, for the first time (so far as we know), that the earth is not the center of the cosmos but merely one of the planets that orbit the sun. By carefully observing the cyclical changes in the moon's appearance as it reflects light from the sun, Aristarchus determined that the sun must be much farther from the earth than is the moon. He then concluded that the sun is much larger than the earth, and therefore more fit to be the center of the cosmos. Aristarchus thus apparently constructed the first sun-centered model of the universe.

About 50 years after the work of Aristarchus, in the third century B.C., the Greek geographer Eratosthenes made the first accurate determination of the circumference of the earth. Eratosthenes learned that at Syene, Egypt, the sun passes directly overhead at noon on the longest day of the year

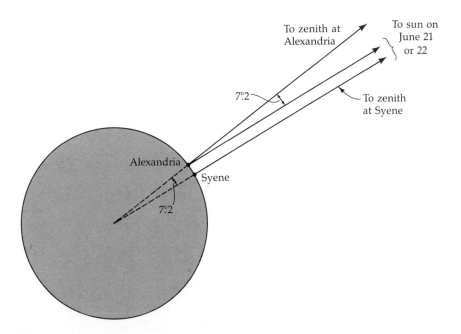

Figure 2-2. Eratosthenes determined the circumference of the earth by measuring the angle between the noon sun and the zenith at Alexandria on midsummer's day, at which time the sun was directly overhead at Syene. This angle equals the angle between the two locations as measured from the center of the earth (which follows from the fact that a line that intersects two parallel lines—the rays of sunlight falling on Syene and Alexandria—makes equal angles with them). The ratio of the distance between Alexandria and Syene to the circumference of the earth equals the ratio of their angular separation (7.2 degrees) to the angular extent of a full circle (360 degrees).

(midsummer's day) and therefore casts no shadow. He also knew that at Alexandria, directly to the north, the sun does cast a shadow at noon on midsummer's day. Eratosthenes concluded from this that the earth's surface must be curved, and that the angle between the noon sun and the zenith (overhead point) at Alexandria on midsummer's day corresponds to the angle between Alexandria and Syene as measured from the earth's center (see Figure 2-2). Eratosthenes hired runners to measure the distance between the two cities, and he correctly concluded that this distance represents 1/50 of the circumference of the earth, since the angle in question is 1/50 of a full circle.

In the second century B.C. lived perhaps the greatest of Greek astronomers, Hipparchus. Drawing on the legacy of observation and intelligent insight of his predecessors over more than four centuries, Hipparchus made several key discoveries about the motion of the planets. He did so despite the fact that he did not accept Aristarchus' model of a sun-centered cosmos but continued to believe in the more popular earth-centered view. The tension between an earth-centered cosmology and one that demoted the earth from its central position was to persist for some 1,800 years.

Hipparchus devised methods to determine the distance from the earth to the moon and to compare the earth-moon distance with the earth-sun distance. He was unable, however, to obtain accurate results. His most impressive achievements dealt with the so-called epicyclic theory of planetary motions.

Observations of the motions of the planets against the background of fixed stars had demonstrated that planetary motions could not be explained as simple circular orbits, traced out at a uniform speed by each planet as it moves around the earth. Today we know that planetary orbits are in fact elliptical and that their center is nowhere near the earth. Hipparchus, however, considered no such possibility, since he was guided by the tradition of the Aristotelian earth-centered cosmology. To obtain agreement with observations while retaining the Aristotelian belief that celestial objects must have perfect circular motions, Hipparchus introduced the idea that the moon, sun, and planets each move along orbits themselves composed of many circular orbits. Superimposed upon each planet's basic circular orbit about the earth were smaller circular orbits that he called epicycles. With this epicyclic theory, Hipparchus was able to obtain relatively good agreement between the observed positions of planets on the sky and the positions he predicted from his model. Hipparchus also compiled a catalog of 850 stars, which, somewhat improved upon by his successor Ptolemy, was the basic reference guide for 1,500 years.

Ptolemy, who lived in the second century A.D. in Alexandria (as did many of his great predecessors), was basically the last representative of the impressive tradition of Greek thinkers. Ptolemy is best known for his refinements of Hipparchus' epicyclic theory. These refinements consisted of further details concerning the number of planetary epicycles and the speeds with which the planets moved along the various epicycles. Apart from his scientific efforts, Ptolemy devoted himself largely to astrology. By the time of Hipparchus, the Romans had established their rule over the Mediterranean world, while leaving intellectual leadership in the hands of the Greeks. The Romans were fascinated by the idea of having one's own horoscope and paid large sums to Greek astrologers, especially when they rendered favorable prognostications.

The Legacy of Greece

Only one Roman, Lucretius, plays a prominent role in the history of cosmological thought. Lucretius, a contemporary of Julius Caesar in the first century B.C., had read all the Greek works on philosophy but was especially influenced by the philosopher Epicurus (third century B.C.), who in turn drew many of his ideas from Democritus. Lucretius expressed his views in a famous six-book epic poem, *De Rerum Natura (On the Nature of Things)*, the beauty and insight of which led to its preservation through the centuries. Here Lucretius expressed his belief that the universe must be infinite. He felt that a finite universe would be imperfect; moreover, he said, the contents of a finite universe would all fall to its center (thus foreshadowing the later reasoning of Newton).

Lucretius' poem is remarkable, too, because it does not hesitate to draw logical conclusions, no matter how startling. For example, having stated that the universe must be infinite, Lucretius infers that it must contain an infinite variety of life. He thus was the first, so far as we know, to assert the existence of inhabited worlds beyond our own.

With the demise of the ancient Greek civilization, Western understanding of the universe stagnated for more than a thousand years. The Roman Empire produced no significant new insights and, after its fall, the Roman Church's suppression of free scientific thought halted further advances in our knowledge throughout most of the Middle Ages. During the period in which the Roman Empire fell into decay, the great library at Alexandria was destroyed, and most of Greek learning was lost or hidden for hundreds of years. Greek cosmology remained basically undiscovered until Arab scholars, during the period from the ninth to thirteenth centuries A.D., finally realized that they had a treasure trove of ancient information at their disposal. These scholars translated many great Greek works into Arabic, and in some cases improved upon them with their own careful observations. Today we refer to Ptolemy's greatest work, the *Syntaxis*, by its Arabic title, *Almagest* ("the greatest").

With the revival of learning in Europe during the Renaissance, the ancient intellectual heritage of the Mediterranean spread northward through Italy to France, Germany, the low countries, and England. Rediscovered Greek classics hinted at realms of knowledge that were far more advanced than those of the Middle Ages. The scientists of Renaissance times typically were churchmen, supported by the faithful and allowed by their superiors to lead a life of scholarly investigation. But the freedom to speculate openly about the nature of the universe was still disallowed by the Church. Among this group of churchmen we find the famous astronomer of the later medieval period, the man who began modern astronomy, Nicolaus Copernicus.

The Copernican Revolution

In the sixteenth century two heretical proposals were introduced to Western science. These ideas bore a curious similarity to some of the concepts of the more radical Greek thinkers (notably Aristarchus and Democritus) almost two millennia before. The first idea was the proposition ad-

Figure 2-3. Copernicus' great work *De Revolutionibus Orbium Coelestium* (*On the Revolutions of the Celestial Orbs*) presented the basic, correct model of the solar system: All the planets orbit the sun. (Photo: Charles Eames.)

vanced by Nicolaus Copernicus that the sun is the central body in the solar system, controlling the motion of the earth and other planets.

Copernicus and the Demotion of the Earth

Copernicus was raised by a prominent uncle to be an educated churchman, and during the last years of the fifteenth century Nicolaus traveled to Italy to study science and medicine. Upon his return to his native Poland, Copernicus became a canon, a minor church official with modest worldly duties. He spent much of his time attempting to understand the models of the cosmos he had encountered in Italy, and eventually came upon the great simplification discovered, or at least imagined, by Aristarchus 1,800 years before.

Copernicus saw that the motions of the planets against the backdrop of the sky could be far more easily explained if the earth is not envisioned as the stationary center of the cosmos. The heliocentric model of the heavens (see Figure 2-3) made far better sense. His model is an important illustration of a principle that still proves successful in discovering the laws that govern the universe: The view of nature that is simplest (mathematically) is likely to be the correct one. Copernicus knew that the weight of tradition and religion lay against this sun-centered model, and he refrained from publishing a full exposition of his theory until he felt himself near death. In 1543, when the future Queen Elizabeth I of England was ten years old, Copernicus' masterwork finally appeared, in an edition of a few hundred copies. These few copies sufficed to break the chains of anthropocentric thought that until this time had enslaved astronomy.

According to the Copernican model, all the planets, including the earth, revolve around the sun. The planets closer to the sun than the earth—Mercury and Venus—move more rapidly in their orbits than does the earth; the planets more distant from the sun—Mars, Jupiter, Saturn (and the subsequently discovered Uranus, Neptune, and Pluto)—move more slowly. This heliocentric model provided a great mental simplification in understanding the workings of nature, but its most obvious difference from Hipparchus' model lay in its dethronement of the earth as the center of the cosmos. Copernicus could not show that any great improvement in accuracy in predicting positions of the planets relative to the stars followed from his model, because he clung to the Greek hypothesis that circular motion is the only natural one for celestial objects. It was left to a later generation of astronomers to show how, with refinement, the model was far superior to any earth-centered model in making such predictions.

The second radical proposal introduced in the sixteenth century was, in a way, even more visionary. Giordano Bruno, an Italian, and Thomas Digges, an Englishman, separately argued, in effect, that the universe has no center. This conclusion followed from their belief that stars are but other suns with their own planetary systems and are scattered throughout infinite space. No particular star, such as the sun, can claim a special position. Bruno was less cautious about revealing his ideas than had been Copernicus. For his efforts, Bruno was burned at the stake in 1600 by the Church in Rome.

Tycho Brahe and the Rise of Observation

In 1546, three years after the death of Copernicus, a man named Tycho Brahe was born. He, like Copernicus, came from northern Europe, which was less intellectually developed than the southern lands to which he journeyed to obtain the best education then available. Tycho learned all the astronomy of his time, but he never accepted the Copernican model of the solar system, although his accurate observations of the motions of the planets eventually demonstrated its validity. What Tycho did, however, had a far-reaching impact on science: He established the value of careful, repeated observations.

In 1572 Tycho observed an apparently new star in the heavens—what we now know to be a "supernova," the explosion of a dying star. For several weeks this supernova was brighter than any other star in the sky, visible even in daytime. Tycho made several observations of the star's position on the sky relative to other stars and showed that the star's position did not change. Hence, he concluded, the new star must be much farther from us than are the planets, because the planets' positions can be seen to change over such a period of time. This observation thus placed this "new star" on the farthest celestial sphere, which, however, according to the Aristotelian tradition, must be eternal and unchanging. The supernova of 1572 thereby destroyed a belief that had prevailed for almost two millennia.

On the island of Hven, the use of which he had obtained by a grant from the king of Denmark, Tycho constructed a castle devoted to his observations of the heavens. Tycho built the best instruments of his time, though no telescopes, which were not invented until the next century. A large wall quadrant enabled him to measure accurately the positions of the stars and planets as they passed across the sky (see Figure 2-4).

Using these instruments, Tycho devoted years to careful observation and equally careful record keeping. Eventually he left Denmark after a row with the new king, and, at the invitation of the Holy Roman emperor, he finally moved to Prague to become "imperial mathematician." There, during the last years of the sixteenth century, a fateful interplay occurred between two extremely different personalities—those of Tycho and a young man who was to overtake him in understanding the cosmos, Johannes Kepler.

Kepler and the Orbits of the Planets

Johannes Kepler, born in 1571 in what is now southwestern Germany, received his education at the University of Tübingen, not far from his family home. Unlike Copernicus and Tycho, Kepler never had the opportunity to travel to the most famous universities, nor did he ever have the chance to meet Galileo Galilei, his contemporary. At Tübingen, Kepler learned of Copernicus' theory and embraced it eagerly. He spent the rest of his life attempting to work out the mathematical "harmony of the worlds," which he imagined must follow from Copernicus' basic idea that the planets orbit the sun. Kepler and his work fall between medieval and modern science, between the old belief that the cosmos reflects a plan of ultimate perfection that can be revealed by pure reason and the modern idea that through patient observation and the use of deductive power, we can advance toward a better understanding of the world around us, without necessarily ever obtaining the ultimate "truth."

Kepler's first attempt to refine the Copernican model used the five "perfect solids," beloved by Greek mathematicians, to explain why exactly six planets orbit the sun (only six were known in Kepler's day) and why the orbits have the relative sizes that they do. This attempt, although inspired,

Figure 2-4. This contemporary engraving shows Tycho Brahe on the island of Hven, directing work at his observatory. The great quadrant measured the altitudes (angle from the horizon) of stars and planets as they crossed the north-south meridian (i.e., when they were directly to the south) while the best clocks available marked the time of meridian passage. (Photo: Jacob Zeitlin.)

ended in failure; and Kepler, in a great display of intellectual honesty, abandoned this theory, much as it attracted him. At the time Kepler proposed this theory, he was an obscure schoolteacher in an unimportant province of Austria. His book *Mysterium Cosmographicum (Mysteries of the Cosmos)* did suffice, however, to bring him to the attention of the astronomers of his day.

Tycho Brahe, seeking an assistant who would help him decipher the mass of data he had collected from his years of observation, wrote to Kepler offering to hire him at an extremely low wage. Kepler hastened to Prague, where he soon found that he was the butt of jokes as a relatively unworldly boor from the provinces. He also found that Tycho was ambivalent about sharing his treasure of accumulated information. Fate intervened in 1601 when Tycho suddenly died, apparently from overindulgence at one of the emperor's banquets. The emperor appointed Kepler to replace Tycho, though at a far lower salary, most of which was never paid even after repeated requests.

In 1609 Kepler deduced from repeated study of observations of the planet Mars the first two of his three laws of planetary motion. First he concluded that Mars must move in an elliptical, rather than a circular, orbit around the sun. He then reasoned that the other planets in the solar system must move in elliptical orbits as well. Kepler's second law of planetary motion states that a planet moving in its orbit around the sun moves more rapidly when it is closer to the sun, in such a way that the imaginary line joining the planet to the sun sweeps over equal areas in equal amounts of time (see Figure 2-5). With these two laws, Kepler swept into obscurity the dependence on perfect circles. He did so only after years of investigating and rejecting other possibilities to explain the motions of the planets, and after arriving at the correct conclusion that Tycho's observations were sufficiently accurate to disprove the hypothesis of epicyclic circular motion that had hitherto prevailed.

In 1619, ten years after Kepler put forth his first two laws, he published his book *Harmonices Mundi (Harmonies of the World)*, in which he announced his third law of planetary motion. This law states the relationship between the time a planet takes to complete one orbit about the sun (its orbital period) and its average distance from the sun: The orbital period of a planet is proportional to the cube of the square root (the $\frac{3}{2}$ power) of its average distance from the sun.

With his three laws of planetary motion, Kepler provided a simple mathematical description of the motions of the planets in the solar system. Kepler, however, could not explain *why* the planets obey the laws he had set forth, although he did suspect that the answer lay in a force produced by the sun.

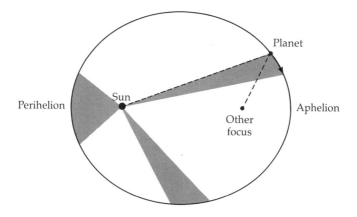

Figure 2-5. Each planet moves along an ellipse, defined as a curve along which the sum of the planet's distances from each of two points (the foci of the ellipse) remains constant. The sun occupies one focus; the other focus is a fixed, imaginary point in space. The planet's speed varies in such a way that the imaginary line joining it to the sun sweeps out equal areas in equal times. Examples of equal areas swept out in equal times are shaded. We see that the planet must move most rapidly at perihelion (when it is closest to the sun) and least rapidly at aphelion (when it is farthest from the sun). The planets' orbits are more nearly circular than shown here.

Galileo and Newton

The man who discovered the nature of the force that holds objects in orbit was Isaac Newton, born in 1642, a dozen years after Kepler's death. Newton's mathematical analysis of nature, using the laws of motion and gravitation that he discovered, marked the beginning of modern science.

Newton's breakthroughs in revealing the mathematical laws that govern the universe were foreshadowed by the contributions of Galileo, who lived from 1564 to 1642. Galileo played a key role in the development of the theory of the motion of objects, but his most important contribution to astronomy was his use of the telescope, invented in Holland a few years before Galileo improved on its basic design. This invention represented the first major extension of human ability to observe the cosmos. The unexpected sights revealed by Galileo's telescopes presaged a recurrent phenomenon: The introduction of a new means of observation to astronomy usually reveals unexpected objects and events. This process, which has accelerated remarkably in recent years, provides the driving force behind our increasing knowledge of the universe. The nature of the cosmos cannot be revealed by theory alone; we must continually seek to deepen our understanding through observation and experiment.

Using telescopes he made himself, Galileo looked at the moon and found ranges of mountains. He looked at Venus, and from its changing illumi-

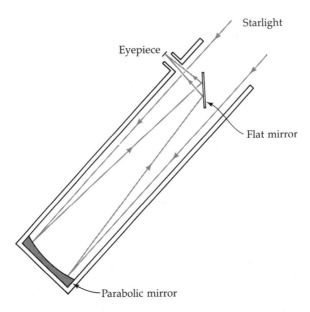

Figure 2-6. Newton's design for a reflecting telescope used a parabolic mirror to focus the incoming light and a smaller, flat mirror to reflect the light to an eyepiece for detection.

nation by the sun showed that it must orbit the sun, not the earth. He looked at Jupiter and found that the giant planet had four satellites. Galileo used his observations of the satellites in orbit around Jupiter to show that their motions follow the same laws that Kepler found to govern the planets (the massive central body being Jupiter rather than the sun). Turning his telescope to the diffuse band of light we call the Milky Way, Galileo showed that this light comes from a multitude of stars too faint to be seen individually with the unaided eye. Galileo was threatened with torture by the Roman Church for his claim, proven correct, that his direct observations further supported the Copernican view that the earth occupies no special position in the universe.

Throughout the hundred years following Galileo's death impressive advances occurred in the making of telescopes. Christian Huygens in Holland improved on the basic refracting telescope design, and Newton invented a new type of telescope, the reflecting telescope (see Figure 2-6). With these telescopes, astronomers made many new discoveries. They found new satellites of the planet Saturn. They discovered many double stars, apparently gravitationally bound to each other. They saw new comets and determined the orbits of these interplanetary wanderers. They observed a number of strange, diffuse objects in the sky, whose appearance differed from the pointlike images of stars, and named them "nebulae," meaning "clouds" in Latin.

These nebulae turned out to include a number of different types of objects. Some were gas clouds, which we now know to lie within our galaxy, and some turned out to be other galaxies, lying far beyond our own Milky Way. To understand the distinction between the two types of nebulae—gas clouds and galaxies—required more than a century of astronomical effort. Only when we discovered that the Milky Way is but one galaxy among multitudes of galaxies was the Copernican Revolution, in its broadest sense, complete. A long history of observations has demoted the earth, then the sun, and finally our own galaxy from any special location in the universe.

The Dark Night Sky

One of the most important observations in the history of cosmology is one that each of us has made, usually without pausing to consider its significance: The sky is dark at night. Although this fact may appear unextraordinary, the darkness of the night sky tells us something profound about the distant reaches of the universe.

The first person, so far as we know, to ponder the significance of this fact was Johannes Kepler. In 1610, when Kepler received a copy of Galileo's book, *Sidereus Nuncius (Starry Messenger)*, he wrote to Galileo saying that he did not believe Galileo's argument that the universe must be infinite, containing an infinite number of stars. If this were so, argued Kepler, then "the celestial vault would be as luminous as the sun." In other words, if we imagine ourselves to be surrounded by an infinite sea of stars, each line of sight (direction of viewing) outward from the earth must eventually terminate on the surface of a star (see Figure 2-7). We may compare the situation to being lost in an infinite forest of bare trees. In whatever direction we look, we see the trunk of a tree at some distance. If we imagine the trees to be shining like stars, we can see that even though most of these shining objects (stars) are quite distant, each line of sight must nonetheless carry a brightness equal to that of the surface of a star, so long as the light has had time to reach us.

In 1610, in his *Conversation with the Starry Messenger,* Kepler published his reply to Galileo's notion of an infinite universe. This work was widely read and no doubt led various astronomers to wonder whether the darkness of the night sky means that the universe is finite. In 1720, the astronomer Edmund Halley proposed that Kepler's objection to an infinite universe could be overcome by assuming that extremely distant stars contribute individual amounts of light too faint to be detected. Halley's argument, however, suffered from a flaw in logic: The fact that an individual star's light may be too faint to be detected does not negate the fact that a

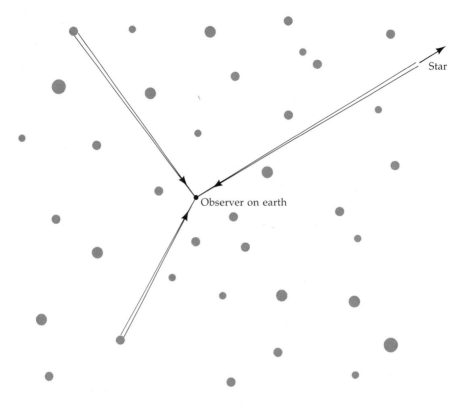

Figure 2-7. If the distribution of stars extends to great distances and stars have always existed, then in every direction we look, our line of sight must eventually encounter the surface of a star. Here the size of stars relative to their separation is greatly exaggerated, but this does not affect the conclusion.

huge number of stars will collectively contribute the large amount of light expected. Similarly, the fact that our eyes are not sufficiently sensitive to detect the light emitted by a single atom does not prevent us from seeing the objects around us.

In 1744, two dozen years after Halley presented his counterargument, the Swiss astronomer Jean Philippe de Cheseaux suggested that matter between the stars, in interstellar space, absorbs starlight, thus explaining how an infinite universe might not necessarily produce a night sky as radiant as the sun. De Cheseaux's insight in suggesting the existence of interstellar matter was brilliant, though not to be verified for another century. He was incorrect, however, in suggesting that interstellar absorption of starlight could reduce the brightness of the night sky. When any matter absorbs starlight, it absorbs energy and thus must become hotter. Such heating causes the interstellar matter itself to radiate energy, and given enough time an infinite universe filled with stars and interstellar matter will still produce a night sky as bright as the surface of a star.

The existence of interstellar matter first suggested by de Cheseaux was asserted again in 1826 by Heinrich Olbers. Forgetting Olbers' distinguished predecessors who also recognized this problem, astronomers often refer to the difficulty in explaining the dark night sky as "Olbers' paradox."

The precise statement of the paradox is this: If the universe were infinite in extent, and filled more or less uniformly with stars having constant luminosity (hence infinite age) and no overall motion, the sky should be as bright as the sun in all directions. To resolve the paradox, we must abandon at least one of the assumptions it contains. Either the universe is *not* infinite in extent or it is *not* sufficiently old or there *is* an overall motion of stars.

As we shall see later, not stars but galaxies of stars are spread more or less uniformly throughout the region of space we have observed, but this does not resolve Olbers' paradox. It is always possible to imagine that a "cosmic edge" to the universe exists, beyond which galaxies of stars become less and less abundant. However, we have no evidence for such a cosmic edge, and we have no reason to believe that our part of the universe differs from any other. In fact, we now know that the other two assumptions that underlie Olbers' paradox are false. Stars do not radiate forever, and the universe is not static (it is expanding).

To understand this resolution of Olbers' paradox, consider just how far you would have to look out into space before your line of sight would reach a stellar surface. Given our knowledge of the average size of a star and the distribution of stars in space, we can calculate that a typical line of sight would extend almost a trillion trillion (10^{24})* light-years from earth. A light-year, the distance light travels during one year, is comparable to the distance from earth to the nearest stars (excluding the sun).

Because each star contains only a finite amount of available nuclear energy, it can radiate that energy for only a finite time, about 10 billion (10^{10}) years, as we shall see in Chapters 4 and 5. Furthermore, we believe that the universe has existed, at least in its present form, for no more than about 20 billion years. Hence, even if the universe is infinite in extent, we cannot hope to receive light from stars more distant than about 20 billion light-years. (Even if we could, the red shift of their light produced by the expansion of the universe [see Chapter 6] would greatly reduce the brightness of the more distant stars.) To put this another way, the sphere around the earth from which we can hope to receive any starlight has a radius equal to only about one part in 100 trillion (10^{14}) of the radius needed to make the night sky as bright as a stellar surface. Here lies the resolution

---

*10^n represents 1 followed by n zeros ($10^5 = 100,000$); 10^{-n} represents 1 divided by 10^n ($10^{-5} = 1/100,000$).

of Olbers' paradox: The universe is far too young for us to expect the night sky to blaze with light.

Astrophysicist Edward Harrison, who has analyzed Olbers' paradox in detail, also discovered that Edgar Allan Poe apparently was on the right track in 1848 (the year before his death). Poe wrote, "The only mode, therefore, in which . . . we could comprehend the voids which our telescopes find in innumerable directions would be by supposing the distance of the invisible background so immense that no ray from it has yet been able to reach us at all."

The Discovery of Other Galaxies

For three centuries after Galileo discovered the nature of the Milky Way in 1609, many astronomers thought of our galaxy as comprising the entire universe. The Milky Way was believed (correctly) to be a pancakelike distribution of stars (one that appears to be a band of stars as seen by us from within the galaxy). However, many scientists refused to relinquish a preferred place for humanity in the universe, and so believed that the sun lay at the center of this disk of stars.

During the nineteenth century, as astronomers built increasingly more sophisticated telescopes, they grew steadily more interested in the "nebulae" they observed, which they noted were of different types. Some nebulae were clearly clouds of gas, lit from within by bright stars. Perhaps the best known of these is the Orion Nebula, which is seen as the middle "star" of Orion's sword. But the nature of other kinds of nebulae was not so evident. Some of these, the "spiral nebulae," had a very different structure (see Figure 2-8).

As the twentieth century began, the debate among astronomers over the nature of the spiral nebulae grew. Some astronomers felt that they were (as we now say) galaxies on their own, similar to our Milky Way but outside it. With remarkable prescience, this view had been put forward more than a century earlier by the philosopher Immanuel Kant, who called such nebulae "island universes." Other astronomers thought the spiral nebulae were merely constituents of the Milky Way, which, they believed, contained all visible objects.

In 1920, before assembled members of the National Academy of Sciences, two distinguished astronomers, Heber Curtis and Harlow Shapley, presented their views about the distances of the spiral nebulae from earth. At that time, methods of estimating the distances of faraway objects were only in the early stages of development. Even today, as we shall see, the determination of the distances to remote astronomical objects is beset with difficulties.

Shapley had recently shown that the sun lies far from the center of the Milky Way, because the distribution of so-called "globular star clusters" in

Figure 2-8. (Top) William Parsons (later the Earl of Rosse) used what was in the 1840s the world's largest telescope to draw the appearance of the "Whirlpool Nebula," seen in the direction of the constellation Canes Venatici. (Bottom) Today we know this "nebula" to be a giant spiral galaxy, seen almost face on. (Photo: Lick Observatory.)

space centers around a point tens of thousands of light-years from the sun. The center of the globular clusters' distribution marks the center of our galaxy. Shapley's figure for the size of the Milky Way turned out to be larger than its actual size, because he neglected to account fully for the effects of the interstellar absorption of starlight. (The absorption of starlight by interstellar matter reduces the observed brightnesses of stars, making them appear more distant than they in fact are.) Shapley therefore believed that the immense Milky Way could contain the spiral nebulae.

Curtis, on the other hand, claimed that his observations demonstrated that the spiral nebulae must be farther from us than the span of the Milky Way. Although in 1920 Curtis could not produce any hard evidence for this conclusion, three years later Edwin Hubble, observing with the new 2.5-meter (100-inch) telescope at the Mount Wilson Observatory, showed that variable stars exist in the most prominent spiral nebulae, and that these stars fluctuate in brightness in a manner identical to those known to be part of our Milky Way star system.

Hubble assumed that the variable stars in the Andromeda spiral nebula have the same intrinsic brightness, or luminosity, as those known to be part of the Milky Way. He could then show, from a comparison of the *apparent* brightnesses of the variable stars, that the Andromeda Nebula must be much farther from us than even Shapley's estimate for the diameter of the Milky Way system. In this way, Hubble produced definite proof that our Milky Way is only one galaxy among many.

The word galaxy (from the Greek word *gala*, "milk"), rather than the word nebula, is now used by astronomers to describe collections of billions of stars, found (as far as we can see) throughout the universe. Our Milky Way, among the largest of galaxies, contains several hundred billion stars, as does the Andromeda galaxy, our closest neighbor similar to the Milky Way (see Figure 2-9). During the first third of this century, as astronomers continued to observe more and more galaxies in greater detail, other types of galaxies besides spirals were discovered. Nevertheless, it became evident that the Milky Way is but a typical galaxy in the cosmic sea of galaxies. This realization changed our conception of the universe in a profound way, laying the foundation for the view of the universe as "uniform." This uniform model, which we shall explore in succeeding chapters, has—after two millennia of scientific progress—replaced the models centered on the earth, the sun, or our own galaxy.

Evolution of Thought and Evolution of the Universe

Within ten years after the spiral nebulae had been shown to be distant galaxies, the work of Vesto Slipher, Edwin Hubble, and their colleagues in determining the motions and distances of galaxies revealed something totally unexpected: The galaxies are moving away from each other, with

Figure 2-9. The Andromeda galaxy, 2 million light-years away, contains about 300 billion stars. Most of these stars have changed little since they emitted the light recorded on this photograph—about the time our ancestors foraged for roots in the Olduvai Gorge. (Photo: Lick Observatory.)

velocities proportional to their separations. The universe is expanding. An expanding universe implies that all the matter was much closer together in the past and will be more dispersed in the future.

The discovery of the microwave background radiation in 1965 (see Chapter 6) provided direct proof that the universe was once much denser and hotter than it is now, because such radiation could have been produced only under those conditions. This discovery came amidst three recent decades of astronomical progress, a "golden era" of astronomy. During the 1950s, astronomers mapped the skies at radio wavelengths. The 1960s brought the advent of X-ray astronomy, with detectors sent above the earth's atmosphere. In the 1970s, the remaining types of electromagnetic radiation—infrared, ultraviolet, and gamma rays (the highest energy form)—joined radio, microwave, X rays, and visible light as accessible to astronomical study. In each new wavelength domain, new types of objects have been revealed by the radiation they emit.

These discoveries have radically altered our conception of the universe—from a universe that is timeless and unchanging to one that is dynamic and evolving. We find a curious contrast here between our past and present notions of the universe's variation *in time* and its variation *in space*. Early cosmologies pictured a universe whose properties changed with distance from us, but were unchanging over time. Now we believe that the universe would look about the same to any observer living in any galaxy, but that its appearance changes with time.

Will our present picture of the universe seem as naïve to future generations as ancient cosmologies appear to us now? This is possible, but we would argue it is unlikely. Our idea of the nature of the universe no longer rests primarily on our *conceptions*, what we think the universe *should be*; but rather on our *perceptions* (enhanced by better and better instruments), what the universe *actually is*. Instead of conceiving, we are perceiving the universe.

The facts that we have gained through reliable observations can never be overthrown in the future; only our interpretation of them may change. Scientists believe, and history has shown, that the steady accumulation of data leads us closer and closer to "reality," an accurate description of the universe. Hand in hand with our progress in observing the universe in more detail and in new ways must come increasingly refined theories, capable of organizing the flood of data into a comprehensive framework of understanding. This framework is based upon the laws that describe how matter behaves, which we shall discuss in the next chapter. The gods controlled the universe of the ancients; the laws of physics control ours.

3

To See a World in a Grain of Sand

To see a world in a grain of sand
And a heaven in a wild flower,
Hold infinity in the palm of your hand
And eternity in an hour.
        William Blake, *Auguries of Innocence*

IN ORDER TO UNDERSTAND THE UNIVERSE, we must understand the laws that govern it. These laws consist of mathematical theories. Progress in our understanding rests upon the predictive power of these theories of nature. A theory is scientifically useless if it only explains results after the fact— it must stick its neck out and expose itself to testing. We can develop confidence only in those theories that have correctly predicted the results of many experiments before they were performed and incorrectly predicted none. If we ask how well certain "theories" that claim to be scientific, such as astrology or biorhythms, adhere to this requirement, we find that they typically fail.

It is important to remember that a theory can only be disproven, never proven. At any time, other theories in existence may also agree with all experiments that have been performed. We can be almost certain that today's favorite theories will prove to be inadequate in the future as experiments probe more of these theories' consequences. A law of nature is not dogma but a guide to understanding, always subject to modification. (Again, the contrast with pseudoscientific theories appears: Few astrologers talk of how astrology could be disproven.)

The Simplicity of Nature

When we look at the characteristics of the presently accepted laws of physics, a remarkable fact emerges: They are simple. This does not mean that the laws are easy to understand or their consequences easy to determine. Although the interaction between two or three particles may be easy to calculate, the behavior of systems containing many particles cannot be determined with mathematical precision. But this limitation arises only as a consequence of the large number of particles involved and could be overcome by a sufficiently powerful computer. We could use these laws to uncover the secret of how the human brain "works" (for instance, how it produces the theories that lead to these laws) if the equations that describe this system of a billion billion billion (10^{27}) atoms could be solved. The marvelous complexity of our world reflects not the simplicity of the underlying laws but rather the many possible structures allowed by those laws.

In what sense are the laws simple? They are simple when described in an appropriate way. And that way is by mathematics—the language of nature. The set of equations expressing the laws can be written down in a few lines, and yet they govern everything. The chief triumph of science has been the discovery of this underlying simplicity of nature.

Another aspect of this simplicity concerns the forms of matter governed by these laws: the elementary particles. At present there appear to be only three distinct classes of particles, with each class containing only a few different particles. Even the forces between all particles are produced by particles of one particular class.

At present no incontrovertible evidence exists to refute the belief that all aspects of nature can be understood in terms of a few elementary particles. It appears that Democritus was right in saying, "In reality there are only atoms and the void."

The Building Blocks of Nature

What do we mean by an "elementary" particle? In essence, we mean a particle with no internal structure and no intrinsic size. But how do we tell whether incredibly small objects really are elementary, or whether they are instead composed of other particles? The most direct way is to study the debris from high-energy collisions between particles, with the collision energy supplied either by nature (in cosmic rays) or by the power company (in particle accelerators).

This technique is analogous to studying the results of an automobile collision in order to determine what comprises an automobile. Of course, we find it far more convenient to study automobiles simply by looking inside them, but this is precisely what we cannot do with matter of the smallest range of sizes. Our picture of matter at these subatomic dimensions is blurred. The Uncertainty Principle of quantum mechanics tells us that the scale of the blurring (the resolution of our picture) is inversely proportional to the momentum of the colliding particles. Hence the greater

Figure 3-1. Electrons and their antiparticles (positrons) are accelerated to velocities nearly equal to the speed of light along the 3-kilometer length of the Stanford Linear Accelerator. Magnets then guide them into counter-rotating orbits within the underground circular section (foreground) where they collide at high energy, producing a great variety of other particles. (Photo: Stanford Linear Accelerator Center.)

the energy of the collision (which implies higher momentum), the smaller the distances that we can investigate. For this reason, scientists continue to build more energetic particle accelerators, such as the one shown in Figure 3-1. Using these instruments, physicists have discovered that particles such as neutrons and protons, once thought to be elementary, are in reality made of other particles, which now replace their predecessors on the list of elementary particles. Furthermore, these accelerator experi-

Table 3-1: Elementary Particles

Particle	Spin	Charge	Mass (MeV)*	Stable?
Leptons (do not feel strong interaction)				
e (electron)	½	-1	0.511	yes
ν_e (electron neutrino)	½	0	$<10^{-5}$	yes?
μ (muon)	½	-1	105.7	no
ν_μ (muon neutrino)	½	0	<0.65	yes?
τ (tau)	½	-1	1,780	no
ν_τ (tau neutrino)	½	0	<250	yes?
Quarks (each comes in three "colors")				
u (up)	½	⅔	4?	yes
d (down)	½	$-⅓$	8?	in protons
c (charm)	½	⅔	1,150?	no
s (strange)	½	$-⅓$	150?	no
t (top)	½	⅔	?	no
b (bottom)	½	$-⅓$	4,500?	no
Bosons (transmit interactions)				
γ (photon)	1	0	0	yes
W	1	-1	80,000?	no
Z	1	0	90,000?	no
gluon (8 types)	1	0	0?	no
graviton	2	0	0	yes

General note: In addition to these elementary particles are their corresponding antiparticles, which have the same mass and spin but opposite electric charge. Antiparticles are denoted either by a superscript indicating their charge (such as a positron e^+, the positively charged antiparticle of the electron e [also denoted by e^-]) or by a bar (such as $\bar{u}$, the antiparticle of the up quark u).

*One MeV is 1 million "electron volts" of energy, equivalent (by Einstein's formula $E = mc^2$) to a mass of 1.78×10^{-27} grams. The symbol $<$ means less than; $>$ means greater than.

ments have guided the formulation of the laws that govern these newly discovered particles.

This process of discovery of an ever-deeper underlying structure might well continue if higher and higher energy probes could be used. However, the size and cost of an accelerator increase as the energy it can impart to particles increases. We may be approaching the practical limit of conventional accelerators. Still, as we shall see, there are other ways to study the workings of nature at the smallest scales.

At a fundamental level, the question of what the world is made of is answered in Table 3-1, the present list of elementary particles. As mentioned earlier, these particles are grouped into three classes: *leptons, quarks,* and *bosons.* For each particle, the table lists the spin (divided by the basic unit of spin, called Planck's constant [$\hbar$]), the electric charge (divided by the charge of a proton), and the mass (in terms of an equivalent amount of energy, in units of one million electron volts [MeV]). The spin measures the particle's intrinsic rotation, which like its charge exists in multiples of a fundamental unit. The table also indicates which particles are stable over the "age of the universe" (about 10 billion years). Unstable particles decay into other elementary particles, usually after a very short time.

The *leptons* (from the Greek *leptos,* meaning slender or light) are so named because the first two discovered, the electron and its neutrino, were much lighter than the other major particles then known, the proton and neutron. (Each proton and neutron has a mass about 1,840 times that of an electron.) Since then, physicists have discovered two other kinds of leptons, the muon and tau lepton, each accompanied by a new kind of neutrino. Notice that the charged leptons are much more massive than their neutrinos, for which we have only upper limits to the mass. Neutrinos interact so weakly with matter that they easily pass through the earth instead of colliding with any of the earth's particles.

The next category of particles, the *quarks,* owes its name to the phrase "three quarks for Muster Mark" in James Joyce's *Finnegans Wake.* Murray Gell-Mann, who (along with George Zweig) first proposed their existence, chose the name because three quarks are needed to make a proton or neutron. This can be seen from the fact that quark charges come in multiples of ⅓ of the charge of a proton (or electron). Although we now have good reason to believe that protons, neutrons, and related particles (called baryons and mesons) are made from quarks, no one has yet seen individual quarks liberated through the collisions of such particles. That is why their masses are not well defined, as is indicated in Table 3-1. Although one experimental group claims to have detected fractional charges in ordinary matter, no other group has duplicated these results (see Chapter 8). The presently accepted theory—called quantum chromodynamics or QCD—which governs quarks posits that quarks interact with each other so strongly that they in fact can never be separated from one another.

The theory of quantum chromodynamics derives its name from the fact that each type of quark—up, down, charm, strange, top, and bottom—possesses a characteristic called "color." The attribute of color has nothing to do with the conventional meaning of the word, but is instead a type of charge (different from the quark's electric charge). Unlike electric charge, the color charge of a quark can assume one of three possible values (e.g., red, yellow, or blue in physicists' language). Like the leptons, quarks seem to come in pairs, which differ mainly in their masses (see Table 3-1). We have discovered the more massive pairs of particles as accelerators have become increasingly energetic. (The top quark has yet to be detected, so if it exists, it must be more massive than the bottom quark.) All but the

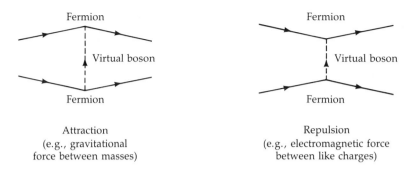

Figure 3-2. The exchange of virtual bosons between particles produces all the forces found in nature. Arrows indicate the direction of motion of the particles. Notice that particles (in this case fermions) can either be drawn together (attraction) or pushed apart (repulsion) by this process.

lightest particles of each type do not exist under normal conditions because these heavier particles are unstable.

The leptons and quarks all have spin ½ and as a group are called *fermions*, in honor of the Italian physicist Enrico Fermi. In contrast, the *bosons* (named in honor of the Indian physicist Satyendra Bose) all have integer spins. More importantly, bosons are the particles that transmit the forces of nature. We can picture the interaction between any two particles as arising from the exchange of what are called "virtual" (not separately detectable) bosons between the two particles (see Figure 3-2). The distance over which any force can act is inversely proportional to the mass of the boson that produces it.

The familiar long-range forces of nature—gravitation and electromagnetism—are produced by the exchange of massless bosons called gravitons and photons, respectively. Gravitation and electromagnetism therefore can act over an infinite range, but their strength decreases with distance. The strong force that binds the quarks together is produced by bosons called gluons, whereas the "massive bosons" (Z [or Z^0, since it has zero charge], W^-, and its antiparticle, W^+) produce the weak interactions responsible for the decay of certain nuclei and other particles. These two forces—strong (also called "color") and weak—effectively act over only extremely short distances: one ten-trillionth (10^{-13}) of a centimeter, the size of a proton or neutron, in the case of the strong (color) interaction carried by gluons, and a thousand times less for the weak interaction carried by the massive bosons. Free (individually detectable) bosons as well as virtual bosons can also exist. Gravitational and electromagnetic radiation represent large numbers of free gravitons and photons, respectively, streaming through space at the speed of light.

No individual graviton has ever been observed (nor will be for the foreseeable future) because gravitons interact so feebly with matter. Yet we have all observed the effect of exchanges of virtual gravitons that produce the gravitational force. Nor have the carriers of weak interaction, the W and Z particles, yet been detected. One of the major goals of physicists in building new accelerators is the discovery of these particles, predicted to have the masses indicated in Table 3-1. Like any other particle, W and Z bosons can be produced when the total energy of two other colliding particles (of appropriate type) equals at least the sought-after particle's mass times the square of c, the speed of light (Einstein's famous mass-energy equivalence formula, $E = mc^2$). Since the W and Z particles interact with other particles much more readily than do gravitons, they should be detectable, even though they live for such a short time that they cannot be seen directly. The products of their decay can provide the signature of their existence (see Figure 3-3).

In addition to the particles listed in Table 3-1 are their partners, the antiparticles. Antiparticles are the "mirror images" of particles, with the same mass but opposite sign of electric charge. (See general note in Table 3-1. The graviton, photon, and Z are their own antiparticles.) When a particle collides with its antiparticle, the pair can disappear, converting its mass and energy into other types of particles. We shall see some examples of this process of annihilation later.

Forces and Motion

Let us now examine some of the characteristics of the four fundamental forces that govern the cosmos: (1) gravitation, (2) electromagnetism, (3) weak, and (4) strong. We begin with *gravitation*, because it is the only universal interaction, affecting all forms of mass and energy. Hence you exert a gravitational force on every bit of matter and energy in the universe, and every bit of matter and energy likewise exerts a gravitational force upon you. Gravitation also has special importance in cosmology because it controls the motion of large aggregates of matter. The force that holds you to the earth is the same force that holds stars and galaxies together and governs the expansion of the universe. In 1687 Isaac Newton set forth this universal behavior of gravitation and the first form of the law of this force in his *Philosophiae Naturalis Principia Mathematica (Mathematical Principles of Natural Philosophy)*, one of the most significant books ever written. Newton was the first person to exploit fully the power of the scientific method. Let us see how he applied it to gravitation.

Newton proposed that the gravitational force between two objects is always attractive, with a strength proportional to the product of the masses of the two objects and inversely proportional to the square of their separation. Newton's great realization was that his theory should apply to the

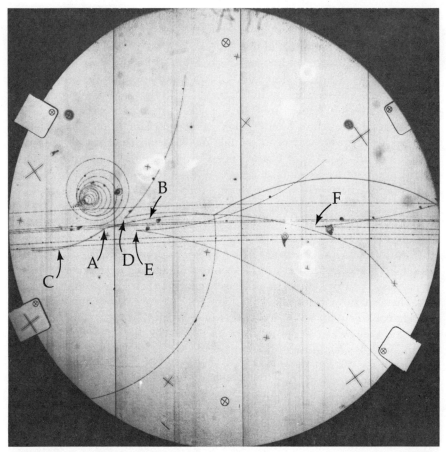

Figure 3-3. This is a photograph of tracks produced by charged particles in a liquid-hydrogen bubble chamber. Each track is composed of many tiny bubbles produced by the passage of the particle through the liquid. A beam of six positively charged π mesons entered from the left, producing the six parallel tracks. At point A one of the mesons collided with a hydrogen nucleus (proton) in the chamber. Two charged particles (B and C) and three uncharged particles were produced by the collision. The uncharged particles (invisible) each later decayed to two charged particles at points D, E, and F. The spiral track was caused by the motion of an electron (ejected from a hydrogen atom by one of the other mesons in the beam) in the strong magnetic field that exists in the chamber. (Photo: Stanford Linear Accelerator Center.)

motion of the moon and all other celestial objects as well as to objects on the surface of the earth. Therefore his theory could be tested by comparing the motion of the moon, for example, with the motion of an object falling toward the earth. Figure 3-4 shows this method of testing.

In addition to a law of force, Newton needed a law of motion to describe how an object responds to a force. The laws of motion that govern ordinary matter had been quantitatively investigated first by Galileo, but it was Newton who developed them fully. In order to analyze the motion of objects, Newton had to invent a new branch of mathematics as well: the differential calculus. The calculus provided a simple, precise way to describe the continuous changes in an object's position and velocity. Newton used this method to derive his key law of motion, which states that an object accelerates in the direction of an applied force, at a rate proportional to the strength of the force and inversely proportional to the object's mass. The acceleration of an object is the rate at which its velocity changes with time. Since the velocity prescribes both the object's speed and its direction of motion, a change in either the speed or the direction of motion (or in both) is an acceleration.

Let us apply these laws to the motion of the moon, at a time when the moon is in the position shown in Figure 3-4. If the gravitational force exerted by the earth were suddenly "turned off," the moon would move in a straight line as shown, with a constant velocity, since the absence of force implies zero acceleration. However, because the earth's gravitational

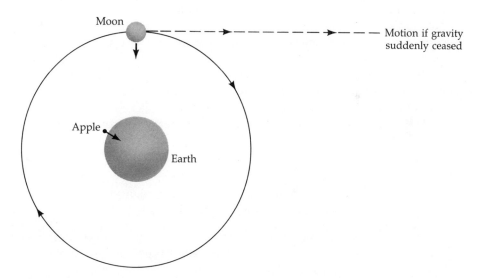

Figure 3-4. The moon deviates from the straight path at constant speed that it would follow if the gravitational field of the earth could be "turned off" suddenly. Because of the acceleration produced by the gravitational field of the earth, the moon falls *around* the earth. Because the apple has no initial velocity, it falls *toward* the earth's center. Arrows toward the center of the earth indicate the direction of the acceleration produced by the gravitational force of the earth.

force cannot be turned off, the moon does accelerate: It "falls" toward the earth, deviating from straight-line motion. (The same analysis applies to the moon's motion at any other time, resulting in the nearly circular orbit shown in the drawing.)

Now consider the motion of an apple dropped near the surface of the earth. The apple, like the moon, is accelerated (toward the ground) by the same gravitational force produced by the earth. Newton asked the following question: What should be the ratio of the distances that the apple and the moon fall during a given interval of time (say, one second)? This ratio should be equal to the ratio of the objects' accelerations. Their accelerations are proportional to the gravitational force they feel divided by the objects' masses. Using his gravitational-force law, Newton then concluded that the ratio of the accelerations—and therefore of the distances the moon and the apple fall during one second—must be inversely proportional to the squares of the distances of the objects from the center of the earth.

From the known properties of the moon's orbit, Newton knew how far the moon deviates from a straight line ("falls") during one second. Knowing the distances of the moon and the apple from the center of the earth, Newton used his formula to predict how far the apple would fall in one second. His prediction agreed with the results of the experiment he performed: The law of gravitation had passed its first test.

Newton's laws of motion and gravitation continued to predict the behavior of astronomical objects correctly for more than two centuries, until superseded by Albert Einstein's theories of special and general relativity. The laws developed by Einstein predict many new phenomena in situations where particle velocities approach the speed of light or where the gravitational force becomes immensely strong. But it is still true that many aspects of the universe can be understood within the approximate framework provided by Newton's laws. The impact of Newton's revelations, which showed that we could truly begin to understand the workings of the cosmos, was beautifully expressed by Alexander Pope:

> Nature and Nature's laws lay hid in night:
> God said, Let Newton be! and all was light.

The second type of force, *electromagnetism,* can under many circumstances be thought of as composed of two different types of forces, electric and magnetic. Fixed electric charges produce electric forces, and moving charges (currents) produce magnetic forces. Precisely because what is fixed and what is moving depends upon the motion of the observer, electric and magnetic forces must be considered dual aspects of a single force, electromagnetism. This unification was achieved within Einstein's theory of special relativity in 1905.

The electric force between two charged particles resembles the gravitational force: It is proportional to the product of the charges on each particle and inversely proportional to the square of their separation. But the electric force differs from the gravitational force because charge can be either positive or negative, so the electric force can be either attractive or repulsive, depending on the sign of the product of the two charges. Mass is always "positive"; hence gravitation is always attractive. Like charges, however, repel and unlike charges attract. This explains why gravitation, not electromagnetism, provides the dominant force between large objects, even though the electromagnetic force between charged elementary particles tremendously exceeds—by a factor of about a thousand trillion trillion trillion (10^{39})—the gravitational force. Gravitation's dominance arises because any charged object tends to neutralize itself by attracting charges of the opposite sign. This tendency results in an object with almost no net charge, which therefore exerts little electric force. Because no negative-mass particles exist, gravitation cannot be neutralized in this way—or in any other.

We can illustrate the precise balancing of negative and positive charges, as well as the relative strengths of the gravitational and electric forces, by the following thought experiment. Assume that slightly more positive than negative charges exist within our bodies and in the earth. How large must this charge imbalance be for the repulsive electric force between ourselves and the earth to exceed the attractive gravitational force and hence push us into space? The answer is that for every million billion (10^{15}) charges, only *one* extra positive charge (proton) not neutralized by a negative charge (electron) would be needed to launch us. Fortunately, we (and the earth) are electrically neutral to a much greater degree than one part in 10^{15}.

We cannot describe the other two kinds of force—*weak* and *strong*—in the same way as gravitation and electromagnetism. That these forces act only over extremely short distances means that we must describe them with the equations of quantum mechanics rather than with the equations of classical mechanics that govern the behavior of most large collections of particles (e.g., familiar objects). Quantum mechanics describes correctly the response of matter to the forces of nature under all conditions. However, the response of an object containing many particles is adequately described by the simpler equations of classical mechanics. Although the strong force, transmitted by the gluons, binds quarks together to form particles such as protons and neutrons, the usual concept of force breaks down (for all four types of interactions) when quantum mechanics must be applied at these short distances. At such distances we must analyze the behavior of elementary particles in terms of their individual interactions with one another. In fact, the weak interaction does not bind any particles together, even though it is intrinsically stronger than gravitation.

The fact that gravitational and electromagnetic forces are carried by massless particles means that corresponding "conservation laws" must be obeyed by all processes governed by these interactions. The masslessness of the graviton implies that energy and momentum remain constant, whereas the masslessness of the photon implies conservation of charge. By the "conservation" of some quantity, we mean that the total value of that quantity cannot change with time, so long as the system we are studying remains isolated from external influences. These conservation laws have great power, because they allow us to draw important conclusions about systems of particles, even extremely complex ones. For instance, we know that the energy that a star continuously radiates must be supplied by some source of energy, in this case nuclear fusion. As another example, we know that a reaction between particles can occur only if the sums of the electric charges before and after the reaction are the same. The conservation laws, like the laws of nature from which they are derived, apply on all distance scales, from the interactions among elementary particles to the dynamics of the entire universe.

During the past few years, physics has undergone a major revolution. It now appears that the equations that describe the four types of interactions between particles are not only simple, but simple in the same way. That is, the laws have the same basic structure. The present hope of physicists, which reflects Einstein's earlier goal, is to find that all four kinds of interactions are but different manifestations of one interaction, expressible by a single unified theory.

Indeed two of the four forces, electromagnetic and weak, now appear clearly to be one. This "electroweak" interaction acts differently at low energies (our environment) from the way it acts at high energies (greater than that required to produce the W and Z particles). The unification of the "electroweak" force appears at these high energies, where the carriers of the force (γ, W, Z) become similar. In effect, the distinction between photons (γ) and the W and Z particles arises only because we live in a "low-energy" epoch of the universe. For demonstrating this synthesis, Sheldon Glashow, Abdus Salam, and Steven Weinberg received the 1979 Nobel Prize in physics.

Higher Levels of Structure

How do the diverse structures that form the world around us arise from the few types of elementary particles that are known to exist? To understand the structure of matter, we begin with the simplest composite objects, the neutron and proton, and then consider increasingly more complex levels of structure.

As mentioned earlier, the neutron and proton basically consist of three quarks, which are held together by the exchange of gluons, as shown in

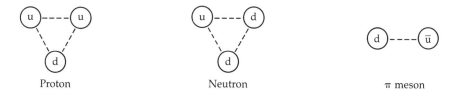

Proton   Neutron   π meson

Figure 3-5. A proton and a neutron each consists of three quarks, bound together by the exchange of virtual gluons (dashed lines). Mesons (in this case a π meson, or pion) consist of a quark and an antiquark (indicated by the bar), bound in the same way.

Figure 3-5. A proton contains two up quarks and one down quark, with the sum of the quark charges equal to the charge of a proton. A neutron, as its name indicates, is electrically neutral (no net charge). It contains two down quarks and one up quark. Other combinations of quarks can also produce particles similar to the neutron and proton, but they all quickly decay into other kinds of particles. A quark can also combine with an antiquark to produce a meson (one type is shown in Figure 3-5); all mesons decay. Other possible combinations of quarks are not allowed by the rule (in QCD) that the color of the quarks must add up to produce an object with no net color. (This is analogous to the case of zero electric charge for the neutron.)

The only stable quark is the up quark. The one exception to this rule is the down quark when it exists within a proton, because a proton cannot decay into other particles. (Actually, no quark may be absolutely stable, as we shall discuss in Chapter 8.) By contrast, after a few minutes, a neutron decays to a proton, an electron, and an antielectron neutrino, as one of its down quarks turns into an up quark.

When protons and neutrons come close enough—to within a distance only slightly greater than their size—they can interact through the same strong (color) interaction that holds them together. This strong force can also hold protons and neutrons together to form nuclei. Furthermore, neutrons cannot decay when they are within most nuclei, because energy would have to be supplied to allow the neutrons to transform into protons. The presence of both neutrons and protons within a nucleus makes possible the rich variety of nuclei that exists in nature.

We characterize a nucleus by its atomic number (Z), the number of protons it contains, and its mass number (A), the number of nucleons (protons plus neutrons) it contains. Ninety-two different elements (nuclei with a given atomic number) exist naturally, with the simplest and most abundant being the hydrogen nucleus, a single proton. Nuclei with the same number of protons (Z) but different numbers of nucleons (A) are called isotopes. Table 3-2 lists the five most abundant nuclei in the universe.

Table 3-2: The Most Abundant Nuclei in the Universe

Element	Symbol*	Z	A	Abundance**
Hydrogen	^{1}H	1	1	90%
Helium	^{4}He	2	4	9%
Carbon	^{12}C	6	12	0.03%
Nitrogen	^{14}N	7	14	0.01%
Oxygen	^{16}O	8	16	0.06%

*The symbol represents ^{A}Z, where Z = the atomic number (number of protons in the nucleus) and A = the mass number (number of protons plus neutrons in the nucleus).
**Abundance is expressed in terms of number of nuclei.

The next level of structure, that of atoms, arises at distances roughly 100,000 (10^5) times greater than the size of a nucleus. Here we find the "orbits" of negatively charged atomic electrons, held at such distances by their electrical attraction to the positively charged nucleus, in much the same way that the planets are held in their orbits by the gravitational force of the sun. But unlike the orbits of planets in the solar system, the orbits of the electrons are *quantized:* The energy of an "orbiting" electron can have only certain allowed values. Moreover, only two electrons can exist in each allowed orbit. In the absence of external influences, an electron will occupy an orbit of the lowest possible energy. Hence as we add electrons around a nucleus, they populate orbits of higher and higher energy. Only hydrogen (Z = 1) and helium (Z = 2) atoms can have all their electrons in the smallest, lowest-energy orbit (see Figure 3-6).

The fact that only two electrons (spinning in opposite directions) can occupy a single orbital state in an atom is a consequence of the Exclusion Principle: the quantum-mechanical rule that no more than one fermion can exist in any particular state. The state of a particle is specified by the quantized values of its properties, such as spin, energy, and momentum.

Under terrestrial conditions, atoms tend to be electrically neutral, so they acquire the same number of electrons as the number of protons in their nuclei. But at high temperatures and densities, such as those within stars, collisions with photons and free electrons tend to knock electrons out of their orbits, leaving atoms called "ions," each of which possesses fewer electrons than protons.

The next higher level of structure, that of molecules, exists because when brought together, many atoms tend to share their electrons with other atoms. A residual electric force between two atoms can exist even though the atoms are electrically neutral over all. This force equals the sum of the forces between the various charges in each atom. Just as atoms combine to form molecules, molecules themselves can combine with one another to produce intricate structures such as the cells in living creatures. All the complexity of the structure of ordinary matter, including life, is a conse-

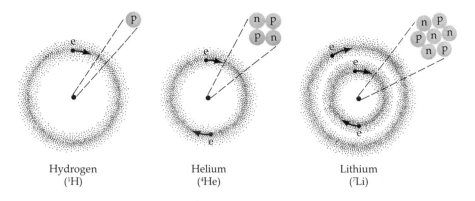

Hydrogen
(^{1}H)

Helium
(^{4}He)

Lithium
(^{7}Li)

Figure 3-6. Hydrogen, helium, and lithium are the three lightest atoms. In any atom, no more than two electrons (e) can occupy the lowest-energy orbit (the one closest to the nucleus). Atomic nuclei consist of protons (p), and, for all nuclei except hydrogen, neutrons (n). The electrons do not actually orbit as shown, since their position at any time cannot be known with certainty.

quence of the electric forces among atoms, or, more precisely, among the constituents of atoms—the electrons and protons—that carry electric charge.

Incidentally, we can now understand why all solids and liquids have roughly the same density (mass per unit volume). For example, we can explain why a cubic meter of concrete has about the same weight as a cubic meter of water. In the solid and liquid states (as opposed to the gaseous state) all atoms are effectively in contact with their neighbors. But the density of an atom—its mass (contained mostly in the nucleus) divided by its volume (determined by the size of its electron cloud)—varies relatively little (by less than a factor of about 100) among all the different types of atoms. The collections of atoms that comprise solids or liquids therefore exhibit a similar relatively narrow range of densities, compared with the almost infinite range of densities possible for gases.

The Special Role of Photons and Gravitons

Almost all of the information we receive from distant regions of the universe comes to us in the form of electromagnetic radiation, which we can picture either as waves or as streams of photons. It is important therefore to understand some of the essential properties of this radiation. First, let us examine some of the characteristics of any massless particle to see why massless particles play such a critical role in astronomy.

The foundation of Einstein's theory of special relativity is the statement that any massless particle (photon, graviton, and possibly neutrino) must travel at a fixed speed, the speed of light. Such a particle's speed through space must therefore be unaffected by the speed of whatever emitted the particle. This requirement leads to many apparent paradoxes in special-relativity theory, but all experiments have verified Einstein's bold prediction. Massless particles travel through the universe in straight lines, unless they are absorbed or scattered by intervening matter or deflected by gravitational fields. By contrast, the other particles that we can detect—the nuclei and electrons in cosmic rays, for example—can have any speed less than the speed of light, but, more importantly, they are deflected by the magnetic fields that exist throughout galaxies because the particles are electrically charged. Hence only with massless particles can we pinpoint the direction of their source simply through observation of the direction from which the particles arrive.

Neutrinos, however, though they may be massless, do not behave quite like photons and gravitons. This is because neutrinos are fermions, whereas photons and gravitons are bosons. The Exclusion Principle allows only a single neutrino to travel through space in a given direction with a given energy. No such restriction holds for bosons, so that great numbers of bosons can travel in the same direction with the same energy. This allows us to describe large collections of identical bosons as coherent waves.

The principal difference between gravitational radiation and electromagnetic radiation is that the accelerated motion of *masses* generates the former whereas the accelerated motion of *charges* generates the latter. Correspondingly, we detect gravitational radiation by the motion it induces in masses, whereas we detect electromagnetic radiation by the motion it induces in charges. (A familiar example of the latter is radio waves, which are generated by powerful alternating currents in radio transmitters and detected by the weak currents the waves induce in radio antennas.)

In addition, because the intrinsic strength of the gravitational force is so much less than that of the electromagnetic force, the detection of gravitational radiation is extremely difficult. Although many experimental groups around the world have sought to detect gravitational radiation, none has yet succeeded. The sensitivities of current experiments, however, are now approaching the level needed to detect the bursts of gravitational waves produced by such violent cosmic events as the collapse of a massive star. In one class of detectors, changes in the 2-meter length of a solid aluminum cylinder induced by a passing gravitational wave can be observed to an accuracy of much less than the diameter of a single nucleus.

We do have indirect evidence for the existence of gravitational radiation. Any system that radiates gravitational waves must lose energy, just as does any system that radiates electromagnetic waves. The principle of conservation of energy demands that this be so. In 1974, radio astronomer Joseph Taylor, now at Princeton University, discovered a pulsar—a rapidly rotating neutron star—in orbit with an unseen companion. By accurately monitoring the regular radio pulses received from the pulsar, Taylor and his colleagues accurately determined the orbit of both the pulsar and its invisible companion. To analyze this system, Taylor had to apply Einstein's theory of general relativity, because in this case Newton's laws do not provide sufficient accuracy.

Through this analysis, Taylor and his colleagues have found that the two objects in this binary star system are slowly spiraling toward each other in their mutual orbit, which shows that the system is losing energy. The rate at which this binary system is observed to be losing energy agrees with the rate at which general-relativity theory predicts that the system should be emitting energy in the form of gravitational radiation. Thus we have observational evidence of the effects of gravitational radiation on the orbit of this binary system, even though we cannot yet detect this radiation directly. Because a binary source such as this produces gravitational radiation of very long wavelength, detection of this radiation would require incredibly accurate monitoring of the relative positions of masses separated by large distances within our solar system. Though the day when we are able to achieve this has not yet arrived, it may occur within our lifetimes.

Let us return now to a property shared by gravitational and electromagnetic radiation, the characteristics of waves. A wave composed of gravitons or photons of a given energy will have a *frequency* (number of vibrations per second) proportional to that energy. The product of frequency and *wavelength*—the distance between successive wave crests—always equals the speed of light (see Figure 3-7). Hence the greater the energy of the particles that form the wave, the higher the frequency and the shorter the wavelength of the wave. Under most circumstances sources emit waves with a broad spectrum of different energies. In the case of electromagnetic radiation, we give different names to the different regions of the spectrum (see Figure 1-4), primarily because of the differences in the way we detect radiation of various wavelengths.

One of the most important ways to analyze the contents of the universe employs the interaction of electromagnetic radiation with matter. Consider a beam of photons impinging on a group of atoms. An electron in an atom can move to an orbit of higher energy if, and only if, the atom absorbs an amount of energy exactly equal to the energy difference between the electron's initial and final orbit. The electron can also drop down to an orbit of lower energy by losing precisely the corresponding amount of energy. In this way, electrons in atoms can gain or lose energy through the absorption or emission of a photon of the required energy.

Hence if we shine light through a collection of atoms, we observe dark lines in the spectrum of the emerging radiation. These dark lines correspond to an absence of those photons that had the correct energies to be absorbed, producing transitions of the electrons to higher energy levels of the atom (see Figure 3-8). By contrast, a hot gas can produce bright lines in the spectrum of its light, the result of photon emission at particular wavelengths as electrons drop into lower energy levels.

Spectral lines, the pattern of dark and bright lines of certain frequencies in a spectrum, form a fingerprint that signals the presence of particular types of atoms. The wavelengths at which the spectral lines appear differ among elements and also depend upon the degree of ionization (the number of electrons lost) of the atoms of each element. These spectral lines play a crucial role in astronomy. The strengths of the lines (their relative brightness or darkness compared with the unaffected radiation at nearby frequencies) allow us to determine the abundances of the elements in the particular object we are observing (a star, an interstellar cloud, or even an entire galaxy).

Spectral lines also play a critical role in astronomy because they allow us to determine how fast the object that produced them is moving either toward or away from us. The component of the source's velocity along our line of sight is called its radial velocity. The source's radial velocity changes the wavelengths of the photons we observe. This change is called the

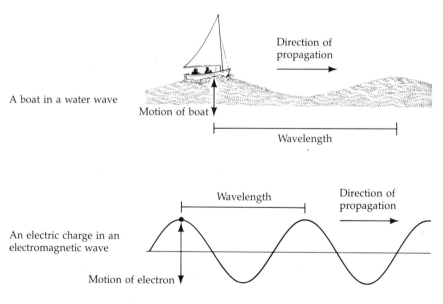

Figure 3-7. Any type of wave motion has a characteristic wavelength and frequency of oscillation. The product of wavelength and frequency equals the speed with which the wave propagates. The motion of a boat on a water wave resembles the motion of an electron (or any other charged particle) in an electromagnetic wave: Both the boat and the particle oscillate up and down as the wave (moving to the right) passes by them.

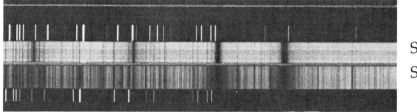

Star

Sun

Figure 3-8. The spectra pictured here (from the sun and from another star) show the intensity of light at different wavelengths. (The spectra above and below the two stellar spectra are reference spectra obtained in the laboratory. Excited atoms produce the bright emission lines.) Both stellar spectra contain dark absorption lines. These lines are produced when photons of certain wavelengths are absorbed by atoms near the surfaces of the two stars. (Photo: Lick Observatory.)

"Doppler effect," after the Austrian mathematician and physicist Christian Doppler, who discovered the relationship from a study of sound waves.

However, the Doppler effect we use in astronomy applies to massless particles, and arises purely as a consequence of special relativity. For radiation reaching us from a source that is moving toward us, the wave crests become compressed because the source has moved closer during each period of oscillation (see Figure 3-9). The time interval between crests is less than it would be if the source were stationary, so motion toward us increases the frequency and decreases the wavelength. Conversely, if the source of waves moves away from us, the waves are "stretched out": The frequency is lower and the wavelength longer (see Figure 3-9). This summarizes the essence of the Doppler effect, which relates the shift of wavelength to the radial velocity of the emitter. (The simple Doppler relationship becomes invalid when the source's velocity approaches the speed of light.) The Doppler effect always depends only on the *relative* motion of the source and observer, since the relative velocity is the only velocity that has any physical meaning.

The fractional shift in frequency or wavelength—as compared with the frequency or wavelength in the case of no relative motion—is proportional to the velocity of the source relative to the observer. If we observe a source that emits photons of particular wavelengths, then larger relative velocities of approach imply shorter wavelengths: These are called "blue shifts." Larger velocities of recession imply longer wavelengths: These are called "red shifts." (Recall that radiation with the longest visible wavelengths appears red, while that with the shortest visible wavelengths appears blue.) By finding the overall shift in wavelength of the characteristic patterns of spectral lines (each pattern produced by a particular type of atom), we can use the Doppler effect to determine the radial velocity of the source.

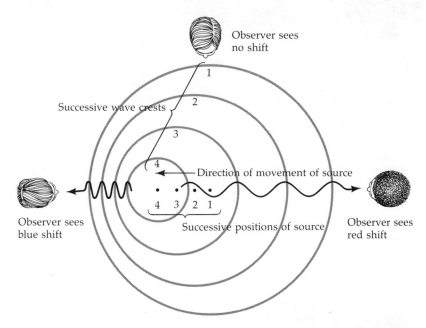

Figure 3-9. Here we see the crests of four successive wavelengths of an electromagnetic wave that has been emitted by a source moving to the left. The numbers label both the position of the source when it emitted each outward-moving wave crest and the present position of that spherical crest. The radiation received by the observer at the left is "blue shifted" because the motion of the source toward her compresses the waves, thus decreasing their wavelength. Radiation received by the observer at the right is "red shifted" because motion of the source away from him increases the separation of wave crests, thus lengthening the wavelength. The observer in the middle sees no shift (unless the source has a speed close to the speed of light).

We shall soon see how such shifted spectral patterns have allowed astronomers to determine the velocities of various objects and to discover the expansion of the universe. But first we shall survey the universe, moving outward from the solar system to discover the diverse ways in which matter populates our cosmic neighborhood.

4

Our Cosmic Neighborhood

WHEN WE SURVEY THE NIGHT SKY, we discover a multitude of objects arrayed over the "celestial sphere." We know now, of course, that these objects do not populate an actual spherical shell, as the ancients thought, but instead span a tremendous range of distances from us. Nevertheless, we can represent the directions in which these objects lie by points on an imaginary celestial sphere surrounding the earth. Although we can measure the directions to these points with great accuracy, our ability to determine the objects' distances is much more limited.

In this chapter, we shall explore the universe in our own cosmic "neighborhood"—a small region in comparison with the size of the visible universe, but still large enough to include the full variety of the forms in which matter is arrayed. This region of the universe lies within a distance of about 1 billion light-years. According to the Copernican Principle, which states that we do not occupy a special position in the universe, the sample of the universe in our neighborhood is typical. The Copernican Principle is supported by the fact that we have not observed any fundamentally new types of objects at distances beyond our neighborhood.

The Constituents of the Universe

All of us have difficulty in developing a feeling for the awesome distances of astronomical objects. Our ordinary units of distance are far too small to be useful. Thus astronomers generally employ as a fundamental unit of distance the light-year, the distance any form of electromagnetic radiation travels in a year: approximately 10 trillion (10^{13}) kilometers.

The Solar System

The moon, our closest astronomical neighbor, is approximately 380,000 kilometers, or about 1¼ light-seconds, from earth. We therefore heard the words of the Apollo astronauts 1¼ seconds after they were spoken on the surface of the moon. The sun is 8.3 light-minutes from earth, whereas Pluto, the most distant planet, is about 5 light-hours away (5.5 billion kilometers). These distances seem enormous in comparison with those we know on earth (for example, the distance from New York to San Francisco is ¹⁄₇₀ of a light-second), but they are virtually infinitesimal in comparison with the distances of objects outside the solar system.

The solar system contains a variety of objects in addition to its star (the sun), 9 planets, and their 48 known satellites (i.e., moons). Smaller chunks of rock range in size from the asteroids, objects much smaller than the largest planetary satellites but which orbit the sun, down to the smallest meteoroids (called meteors when heated by their passage through the earth's atmosphere and meteorites if they survive to reach the earth's surface). Comets are mysterious objects that spend most of their lives in the outer reaches of the solar system. The ones we see happen to have orbits that take them briefly by the sun. They are then partially vaporized by solar radiation and become visible through the sunlight they scatter from the emitted gases. A flux of particles that continually streams from the sun, called the solar wind, sweeps these gases away from the comet, producing its characteristic tail (see Figure 4-1). Finally, a diffuse distribution of grains of dust pervades the solar system.

We now have good reason to believe that all the objects in our solar system—sun, planets, satellites, asteroids, comets, and dust—formed at about the same time, some 4.6 billion years ago, from a rotating, collapsing cloud that spun out a disk of material to form our planetary system. Most of the mass in the solar system, however, remained within the sun. We have no reason to doubt that similar planetary systems exist around many other stars, though we have yet to observe them. We do know from our study of the motions of stars that in such systems, most of the mass must also reside within the central star. Hence in considering the contribution of various constituents to the mass in the universe—a study that plays a central role in cosmology—we can neglect all forms of "debris" in orbit around stars.

To the Stars

As we journey beyond our solar system, we find not complete emptiness but instead a diffuse gas that fills the space between the stars, the interstellar medium. Its density is far less than that of the best vacuum produced in our laboratories. The average distance between individual atoms in the interstellar medium is about 1 centimeter, a million times greater than the distance between atoms in the air we breathe! Conditions in the interstellar gas vary greatly from place to place—from hot, tenuous regions that fill much of interstellar space to localized, denser clouds, some of

Figure 4-1. Particles ejected from the sun, along with the sunlight photons, push material from the nucleus of a comet into a "tail," often millions of kilometers long. The tail points away from the sun as the comet sweeps by it at high velocity. (Photo: Palomar Observatory.)

Figure 4-2. The Horsehead Nebula is a region of our galaxy containing a dust cloud that partially obscures the light emitted from the gas cloud behind it. The gas cloud is irradiated by the bright stars within it. (Photo: Royal Observatory, Edinburgh. Original negative by U.K. Schmidt Telescope Unit.)

them rich in many types of molecules. Through all these forms of interstellar gas pass both magnetic fields and the high-energy electrons and nuclei that we call cosmic rays. In addition, like the interplanetary medium of the solar system, interstellar space is filled with dust grains, each of which contains upwards of a million atoms.

Figure 4-2 is a photograph of a region of the interstellar medium (in the direction of the constellation Orion) that contains a dust cloud called the Horsehead Nebula. The dust fills the lower portion of the picture, as is indicated by the fact that fewer stars can be seen behind it. The central part of the picture shows an emission nebula, which glows with the light emitted by atoms excited by the radiation from the bright star Sigma Orionis. In the lower left we see a reflection nebula, in which atoms in the gas surrounding the stars scatter some of the starlight in our direction.

The first stars we encounter on our journey are at distances of a few light-years (10,000 times greater than the radius of the solar system). As our survey reaches outward to a few hundred light-years, we encounter millions of stars and notice the great variety that exists among them. Important relations among the properties of stars, however, give us clues to their nature. One of the first and most important of these relationships was discovered independently by Ejnar Hertzsprung in Denmark and Henry Norris Russell in America about 70 years ago. They found correlations between a star's surface temperature and its luminosity (energy radiated per unit time). Figure 4-3 shows these relationships.

Any star radiates energy with a great range of photon frequencies, but most of its energy typically appears in the visible and ultraviolet regions

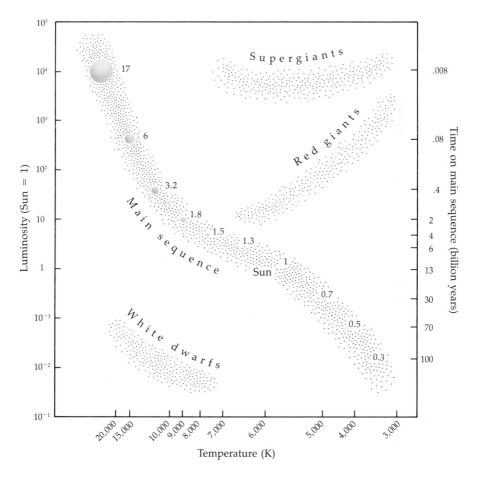

Figure 4-3. The luminosity (compared to that of the sun) and the surface temperature in degrees Kelvin (degrees Celsius above absolute zero) of most observed stars lie in the regions shown on the temperature-luminosity diagram. The diagram also shows the masses of main-sequence stars (relative to the sun's mass) and their lifetimes on the main sequence.

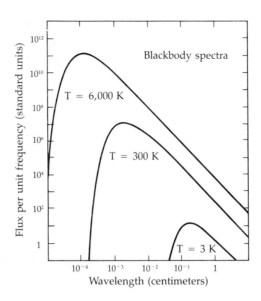

Figure 4-4. Blackbody spectra of electromagnetic radiation characterized by different temperatures have similar shapes. The temperatures indicated (in degrees Kelvin) are those of the cosmic microwave background radiation (3 K), the surface of the earth (300 K), and the surface of the sun (6,000 K). The higher the temperature of a blackbody, the shorter the average wavelength and the higher the average energy of the photons in its spectrum. (Flux is the amount of electromagnetic energy passing through a unit area per unit time. Each division on both axes indicates a factor of ten.)

of the electromagnetic spectrum. A star's spectrum usually contains many dark lines (produced by the atomic absorption of photons, which transfers the electrons to higher energy levels). The general shape of the spectrum resembles that of a so-called blackbody, depicted in Figure 4-4. A blackbody spectrum arises from matter at a uniform temperature that easily absorbs photons of all wavelengths. Such matter produces photons whose intensity distribution at different wavelengths has a characteristic blackbody shape. The total luminosity of an object that radiates like a blackbody varies in proportion to the object's surface area times the fourth power of its surface temperature. The average photon energy (or frequency) varies in direct proportion to the object's surface temperature. A star's surface temperature determines its color, which ranges from blue for the hottest stars to red for the coolest.

A Hertzsprung-Russell diagram such as Figure 4-3 will contain many points, each representing the surface temperature and luminosity of an individual star. These points tend to occupy only certain regions of this diagram. Most of the points form a band running diagonally across the diagram, called the "main sequence." As we shall discuss in the next chapter, most of a star's properties (such as its luminosity and surface temperature) do not change greatly during most of its lifetime. Each point along the main sequence represents a star seen in this quiescent, mid-life phase. A star's location along the main sequence depends primarily on its mass: The more massive the star, the greater the star's luminosity, surface temperature, and radius. The masses of main-sequence stars range from about 100 times the mass of the sun down to $1/10$ of the sun's mass. The less massive stars appear in far greater numbers than do the more massive ones. The sun (a typical main-sequence star) has a mass 330,000 times that of the earth, and its radius equals 2 light-seconds, 100 times that of the earth. Another comparison helps to convey a feeling for the scale of astronomy: The average distance between stars is about 100 million (10^8) times their average diameter.

As stars evolve from their quiescent, main-sequence phase, they grow larger and redder. Stars observed in this phase of their evolution comprise the "red-giant" and "supergiant" regions of the Hertzsprung-Russell diagram. Still later in life stars shrink, and some end their lives as small, dying "white dwarfs," found in the lower-left region of Figure 4-3.

The Structure of Our Galaxy

As we look a few thousand light-years beyond the solar system, we find that stars are no longer evenly distributed over the sky. Instead, the distribution of stars assumes the form of a disk, and we look outward from our position within this disk (see Figure 4-5). As we extend our vision outward tens of thousands of light-years, we discern the full richness of the structure of our galaxy of stars, the Milky Way. The recently formed stars, only a few million years old, highlight the beautiful spiral arms of our galaxy. These stars are the more luminous and massive blue members of the main sequence. Their intense radiation illuminates the interstellar clouds of gas from which they were born (see Figure 4-2, the Horsehead Nebula). The middle-aged stars (a few billion years old) are distributed continuously throughout the disk, whereas the oldest stars (more than 10 billion years old) appear in the halo (the spherical region that surrounds the disk) and the nucleus (the central region of our galaxy).

On a finer scale, more structure appears in the distribution of stars. About half of all stars belong to binary or multiple systems in which two or more stars are held in mutual orbits by their gravitational attraction. Many stars are also bound together by gravitation to form clusters, which contain up to several hundred thousand stars within a region of a hundred light-years or so. Our own galaxy contains about 400 billion stars and spans about 100,000 light-years.

By carefully measuring the shift in the wavelength of the spectral lines of a star, we can determine its velocity along our line of sight because of

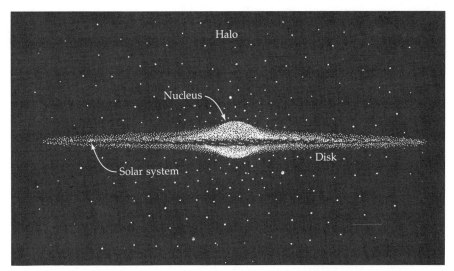

Figure 4-5. An edge-on drawing of our galaxy, the Milky Way, shows the location of the solar system and the distribution of stars within the galaxy.

the Doppler effect, as discussed in Chapter 3. (Recall that the line of sight is the imaginary line from us to the object we are viewing.) Furthermore, we can determine the star's velocity perpendicular to the line of sight by measuring any shift in the star's position on photographs taken many years apart (if the star is close enough to us for its motion to produce a measurable effect).

We can survey only part of our galaxy in this manner, because the interstellar dust grains absorb the light from stars within the disk, thus obscuring our view through the plane of the Milky Way for distances greater than a few thousand light-years. But, as if in cosmic compensation, an electron transition exists in hydrogen atoms (the most abundant element) that produces spectral-line emission and absorption at a radio wavelength, 21 centimeters. Because interstellar grains do not absorb this long-wavelength radiation, we can observe interstellar hydrogen atoms throughout our galaxy (see Figure 4-6).

We find from these observations that the disk of our galaxy is rotating (as one might expect from the pinwheel shape of its spiral arms) once every few hundred million years. This rotation consists of the near-circular orbits of billions of individual stars, which take different amounts of time to make one orbit, depending on their distances from the galactic center. The stars are held in their orbits by gravitation, much as the planets are held in theirs. However, most of the mass does not occupy the center of the system, as it does in the solar system, but is spread throughout the disk and halo, partly in the form of individual stars. The stars in the halo have orbits that pass back and forth through the disk every few hundred million years. The halo stars move like bees in a hive but are always controlled by gravity. Typical speeds for both the disk and the halo stars are a few hundred kilometers per second.

Beyond the Milky Way

As we look beyond the confines of our galaxy, we first encounter the Large and Small Magellanic Clouds, at distances of about 200,000 light-years. These objects, visible only from the Southern Hemisphere, owe their names to Ferdinand Magellan, who noted their existence during his voyage around the world in 1520. The Magellanic Clouds, small satellites of our own galaxy, contain a higher percentage of young stars than does the Milky Way. Unlike our galaxy, which is spiral, the Magellanic Clouds are irregular galaxies; they do not possess the coherent structure found in most galaxies.

At greater distances, we find more small galaxies, whose appearance, however, is much less irregular than that of the Magellanic Clouds. These

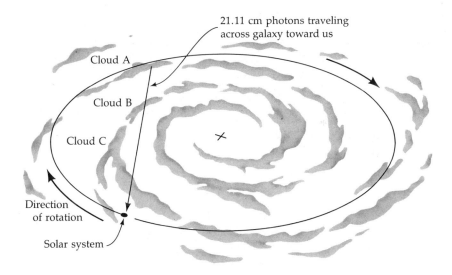

Figure 4-6. We can use the Doppler-shift relationship (shift in wavelength is proportional to velocity) to find the relative velocities along the line of sight of gas clouds in the plane of our galaxy (and in other galaxies). Clouds closer to the center of the galaxy than we are (B and C) move more rapidly in their orbits and are therefore receding from us along the line of sight shown; hence we observe their hydrogen emission at longer wavelengths (red shifted). Cloud A is at the same distance from the center as we are and therefore orbits at the same speed that we do. Because Cloud A has no relative velocity with respect to the solar system, its hydrogen emission (at 21.11 cm) shows no shift in wavelength.

Figure 4-7. The giant elliptical galaxy NGC 4486, also known as M87, contains hundreds of billions of stars, but has less interstellar gas than a spiral galaxy of similar size. (Photo: Kitt Peak National Observatory.)

dwarf elliptical galaxies each contain a few million to a few hundred million stars, spread rather diffusely. Much more massive are the large elliptical galaxies that we encounter at distances of a few million light-years. Such large elliptical galaxies (see Figure 4-7) have masses comparable to the mass of our galaxy, but they do not possess the dominant, rotating disk of gas, dust, and stars that characterizes spiral galaxies. Young stars rarely appear in elliptical galaxies, a fact consistent with the absence of gas and dust in such systems: Previous generations of stars have apparently consumed the stuff from which stars form. Elliptical galaxies range in shape from completely spherical to somewhat flattened systems.

The first galaxy we encounter with a size equal to ours is not an elliptical galaxy. It is Andromeda, a spiral galaxy located at a distance of 2 million light-years (see Figure 2-9). To us, this galaxy looks almost the same as our galaxy would to an observer within the Andromeda spiral.

Clusters of Galaxies

Within 2.5 million light-years, the Milky Way has about two dozen neighbor galaxies. Together these galaxies form a small cluster, called the Local Group of galaxies. Galaxies are held in the cluster only loosely by their mutual gravitational attraction.

As we look beyond the Local Group, we find galaxies spread throughout space, typically separated by about a million light-years. This is the separation of the larger, brighter galaxies, which are the most easily detected and studied. We know that faint galaxies far outnumber bright ones, but most of the mass and luminosity of the galaxies in the universe resides in these larger, brighter galaxies. The two major types of galaxies, spiral and elliptical, are about equally abundant. Irregulars, the third type, comprise only about 10 percent of all galaxies.

Most galaxies, like our own, are not strongly clustered. But we do find some rich clusters of galaxies, containing as many as a few thousand member galaxies (see Figure 4-8). The nearest such cluster, Virgo, is about 50

Figure 4-8. The Hercules cluster of galaxies has several hundred members, some of them larger than the Milky Way. Both spiral and elliptical galaxies are present. This cluster is about 500 million light-years from us. (Photo: Palomar Observatory.)

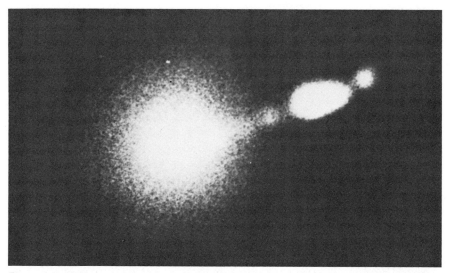

Figure 4-9. A short-exposure photograph of the bright central region of the elliptical galaxy M87 (Figure 4-7) shows a jet of matter that consists of several individual knots. This matter is moving at high velocity away from the galaxy's center. (Photo: Halton Arp, Mount Wilson and Las Campanas Observatories.)

million light-years away. Farther out in space, we find other rich clusters to be distributed throughout the universe in much the same way as in our immediate neighborhood.

Exploding Galaxies

One of the great developments of modern astronomy, still imperfectly assimilated, has been the realization that not all galaxies are as quiescent as they appear. Some galaxies are in the throes of titanic outbursts, expelling matter at enormous velocities from their central regions into intergalactic space (see Figure 4-9). Quite possibly, most galaxies, our own included, pass through such an explosive phase during their evolution. The universe can be violent, with the greatest violence observed in these spectacular galaxies.

Astronomers first became aware of the outbursts of "exploding galaxies" by detecting the enormous amounts of radiation they produce at radio frequencies. These radio waves, generated by the spiraling motion of high-energy charged particles in magnetic fields, come from regions often a million light-years from the galaxy. In many cases two "blobs" of radio emission appear on opposite sides of the galaxy. In other exploding galaxies, radio emission also arises in the central region of the galaxy itself, often, once again, in the form of two blobs (see Figure 4-10). One might

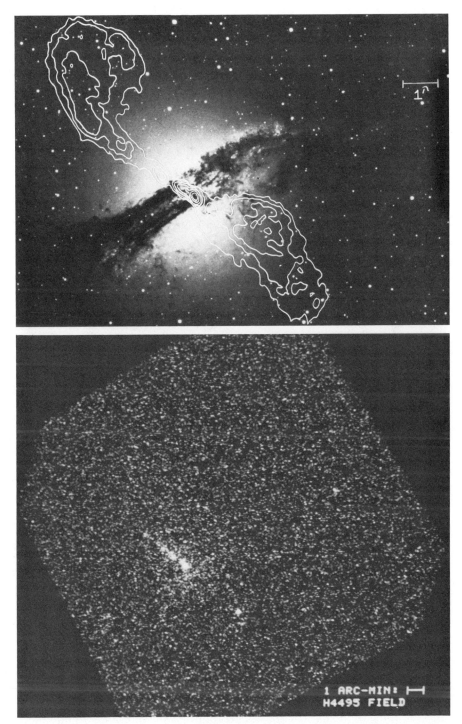

Figure 4-10. Astronomers have mapped the radiation emitted by the exploding galaxy Centaurus A at three wavelengths—visible-light, radio, and X-ray. In the top photograph, contours of equal radio brightness have been superimposed upon the visible-light image. The dark regions on the visible-light photograph result from absorption by dust clouds. (Photo: J.O. Burns, E.D. Feigelson, and E.J. Schreier.) The bottom photograph is the X-ray image of Centaurus A. (Photo: *Astrophysical Journal*, vol. 251, 1981, p. 31.) The bars indicate an angular distance of one minute of arc (1/60 of a degree).

conclude that explosions have ejected matter—the particles that produce the radio emission—in opposite directions away from the core of the galaxy. The amounts of energy involved in these explosions are awesome, reaching a few percent of the total energy radiated by all the stars in a galaxy during the galaxy's entire 10-billion-year lifetime. Yet this energy is released in only a few million years' time.

More surprises lay in store for radio astronomers when they observed in greater detail the regions that appeared to be the origin of the radio "blobs." They found other blobs only light-months apart, yet these blobs were aligned along the same axis as that of the most distant blobs, 10 million times farther away! Apparently the "engine" producing these blobs tends to eject them in opposing directions along the same axis. This axis may be the axis of rotation of the mysterious engine. Additional evidence suggesting the ejection of material came from detailed radio maps, which showed that the innermost blobs are moving apart over a few years' time. Most surprising to astronomers was their discovery of some objects in which the blobs appear to be moving apart faster than the speed of light! This does not violate Einstein's theory of special relativity, since we would have to go to the site of the explosion to measure the *actual* relative velocity of the blobs. However, the observations demonstrate that the material in the blobs must be moving at a speed close to that of light.

Quasars

Astronomers found that many of these outbursts originated in elliptical galaxies. But they also detected radio emission from directions in the sky in which only starlike images appeared (see Figure 4-11). What could these "stars" be? The mystery was resolved by the astronomer Maarten Schmidt, who in 1963 found that these starlike objects have large red shifts. As we shall see in Chapter 6, these large red shifts indicate that the objects must lie at huge distances from us, which in turn means that they must be extremely luminous in order for them to be visible to us at all. Moreover, the objects, called quasistellar objects—*quasars* for short—emit huge amounts of radiation over the entire electromagnetic spectrum. The most luminous quasar produces thousands of times more energy per second than the most luminous galaxy, which concentrates most of its power output into the visible and ultraviolet parts of the spectrum.

It now appears that quasars are located at the centers of certain galaxies. We cannot clearly see a quasar's surrounding galaxy in photographs because of the quasar's great distance and because its great luminosity far outshines the rest of the galaxy. (Scattering of a quasar's light within our atmosphere makes its image as large as that of a distant galaxy.)

We can deduce that the "engine" that powers quasars and exploding galaxies must be incredibly small, considering the amount of energy these

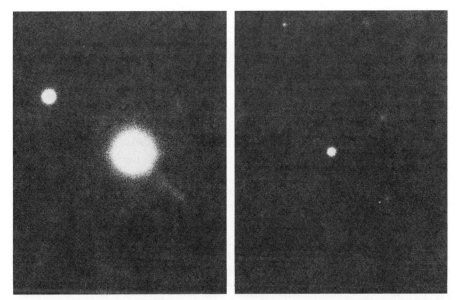

Figure 4-11. Two of the first quasars studied by astronomers Maarten Schmidt and Jesse Greenstein appear as starlike objects in these photographs, although in 3C 273 (left) we see a jet of material apparently ejected from the central quasar. The finite size of the quasars' images is caused by scattering of their light in our atmosphere. These objects are billions of light-years from us. (Photo: Palomar Observatory.)

engines generate. Astronomers determine this by observing that the brightness of some of these apparent galactic cores varies over time scales as short as one day.

Consider what we would see if an object that was much larger in size than 1 light-day flared up for an instant. Light from various parts of the object would reach us at different times. If the object were 1 light-month in size, light from the front of the object would arrive a month before light from the back. Hence we would not see an instantaneous brightening, but instead a relatively slow buildup and gradual decrease in the object's brightness over a period of a month. From this example we can see that no object that emits electromagnetic radiation can appear to vary greatly in brightness in a time *less* than that required for light to cross the object. Thus the engine for the explosive events within galactic cores can be no larger than about 1 light-day, only slightly more than the diameter of the solar system. From this region we see a power output more than a trillion (10^{12}) times that of the sun!

What is this small but powerful engine that produces such jets of high-energy particles and magnetic fields? This question provides one of the great mysteries in astrophysics today. The engine, too small to be seen directly, can be investigated only via mathematical models. One particularly promising theoretical model, which seems capable of explaining much of what is observed, will be discussed in the next chapter.

Almost all quasars lie far beyond our local neighborhood and are thought to be the most distant objects yet observed. However, not all astronomers share the view that quasars lie at such enormous distances. Estimates of the distances to quasars rely on the enormous red shifts observed in their spectra. These red shifts, most astronomers believe, can arise only from the enormous velocities of recession inherent in any object at a correspondingly immense distance from us, since the object must take part in the overall expansion of the universe.

Astronomer Halton Arp of the Mount Wilson and Las Campanas Observatories has led the list of those few who reject this basic assumption concerning quasars' distances. Arp has taken some photographs that show apparent "bridges" of matter between a quasar and a galaxy. Most astronomers believe that the quasar is much more distant and that its position on the sky happens to coincide with the end of the "bridge" of matter emanating from the nearby galaxy. If this were not a superposition arising by chance, it would imply that the quasar and galaxy were relatively close to each other in space (and not only in direction). In that case, the quasar could not have the immense distance suggested by its large red shift, because the galaxy typically has a much smaller red shift and therefore is much closer to us.

For more than ten years, the debate has continued between Arp (and others who reject quasars' distances based on red shifts in their spectra) and those who do assign to quasars the enormous distances their red shifts imply. No firm conclusion has emerged, but the majority view has been strengthened by two types of observations. First, the discovery of quasars in clusters of galaxies with the same red shift indicates that these quasars' red shifts are indeed produced by the expansion of the universe. Second, the images of only the small-red-shift quasars often appear "fuzzy." We can explain this easily if the large-red-shift quasars are at correspondingly large distances, for in such cases the image of the surrounding galaxy (the "fuzz") becomes smaller than the image of the quasar.

Within our cosmic neighborhood, which extends to a distance of a billion light-years from the earth, we believe we have identified the major types of structures that exist throughout the universe. We have found the dominant visible constituent of the universe to be stars. Stars are far from uniformly distributed throughout the universe. They cluster into galaxies, which themselves often form clusters. We shall soon see, however, that at least two other constituents play a critical role in the universe: (1) the microwave background radiation, which is a remnant of the early universe, and (2) some invisible form (or forms) of matter, which we now think provides most of the mass associated with galaxies and clusters of galaxies and thus furnishes most of the mass in the universe.

The Composition of Matter

To complete our local survey of the universe, let us consider the microscopic structure of matter, which contains important clues to its history.

Matter and Antimatter

To begin with, we may naturally wonder whether the universe consists of equal amounts of matter and antimatter, since the familiar laws of physics show no preference for one form over the other. If, for example, every neutron, proton, and electron in the sun were replaced by its corresponding antiparticle, the sun would appear almost the same. We say "almost" for two reasons. First, when the solar wind of antiparticles streaming from the sun reached the earth, it would annihilate with the ordinary atoms in our atmosphere to produce X rays and gamma rays. Second, even if we were far enough from the sun to be unaffected by the solar wind, we could recognize an antimatter sun because the nuclear reactions within the real sun produce predominantly neutrinos, whereas an antimatter sun or antimatter star would radiate predominantly antineutrinos. This method of detection cannot be applied to photons (e.g., sunlight or starlight), for an antiphoton is indistinguishable from a photon.

Does antimatter exist in more distant stars and galaxies? Here we have basically two types of evidence relevant to the question. No antinuclei have been discovered in the cosmic rays that arrive from sources within our galaxy and (for the highest-energy cosmic rays) probably from nearby galaxies as well. This absence of antinuclei provides strong evidence that our neighborhood contains only matter (except for a tiny amount of antimatter that is produced by high-energy collisions and eventually annihilates with ordinary matter). We also know that high-energy photons (X- and gamma-ray radiation) will be produced wherever matter and antimatter come into contact. The absence of clear evidence of such high-energy photon radiation from distant regions of the universe suggests either that our universe consists entirely of matter or that matter and antimatter have been well isolated in separate regions of space. Since we do not see such evidence of annihilation between the galaxies in clusters and the gas through which the galaxies move, these separated regions of matter and antimatter must be at least as large as the clusters of galaxies. Within existing models of the evolution of the early universe, astrophysicists have found no way in which matter and antimatter could have separated on such scales. Thus we appear to live in a universe in which matter far predominates over antimatter, but we do not yet understand why.

Another aspect of the universe's microscopic structure important for our picture of its evolution is the abundance of the elements. (See Table 3-2 for the cosmic abundances of the five most common elements.) We shall see that we can learn still more from studying the abundances of the different isotopes of a given element.

What can we discover about the abundances of elements by analyzing the spectra of light from stars and the interstellar medium within the Milky Way, as well as from other galaxies? First, the oldest stars in our galaxy appear to have formed from gas in which "heavy" elements were far less abundant than in the interstellar medium today. Heavy elements include carbon, nitrogen, oxygen, and the 84 still heavier elements such as neon, silicon, calcium, iron, gold, and uranium. The light elements are hydrogen, helium, lithium, beryllium, and boron. In contrast to the relatively low abundance of the heavy elements in the oldest stars, helium appears to have almost the same abundance in the oldest stars as in younger stars such as our sun. In addition, astronomers have recently discovered entire galaxies that appear generally deficient in heavy elements though they contain a normal abundance of helium. Thus the abundance ratios of the elements in these galaxies resemble those that existed in our galaxy long ago, when the oldest stars were forming.

These facts strongly support the widely held view (backed by detailed calculations of the evolution of stars) that the stars have produced all of the heavy elements in the universe. At the end of their lives, certain massive stars explode, seeding the interstellar medium with the elements they have made. In this way, the interstellar medium has become gradually enriched in heavy elements. This explains why the surfaces of the oldest stars, which represent samples of the ancient interstellar medium from which they condensed, show the lowest abundance of heavy elements. And those galaxies that exhibit an overall deficiency of heavy elements may be young, in the sense that their stars have not yet had time to produce the same relative abundances of heavy elements that we see in typical galaxies.

However, an intriguing fact remains: Not a single star has yet been discovered with *no* heavy elements. Yet the first generation of stars must have formed from gas that contained no heavy elements, if stars make these elements. Where, then, are the stars that long ago enriched the gas from which these oldest visible stars formed? They may have been extremely massive stars that eventually collapsed to form black holes (which we shall discuss in later chapters).

The observations of element abundances in stars also support the view that the bulk of the universe's helium appeared before any of the stars we

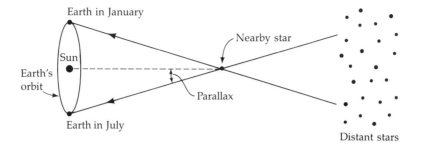

Figure 4-12. The earth's yearly motion around the sun produces a "parallax shift" of the apparent positions of nearby stars relative to much more distant ones. The amount of this shift varies in inverse proportion to each star's distance, but for all stars amounts to less than one second of arc (3,600 seconds of arc = 1 degree of arc). Note: Drawing is not to scale.

see now had formed. According to the prevailing view, helium was made from the neutrons and protons in the enormously hot early universe. It is also possible that the same objects that made the trace of heavy elements seen in the spectra of the oldest stars might have made almost all the helium in the universe as well.

Distance and Knowledge

Let us return to a fundamental issue in our attempt to understand the universe. How do we acquire our knowledge of remote objects? Why does this knowledge become less certain as we look to greater distances? We must be concerned here with the methods that astronomers use to determine many of the properties of our universe and with our prospects for reducing the uncertainties inherent in these methods. We shall confront these questions by considering how the four most basic properties of an object—its distance, luminosity, size, and mass—are determined. Of the four, the distance to an object is of primary importance because we cannot accurately determine most of the object's other properties without knowing its distance.

Distance Measurements by Parallax

Let us now compare some of the common methods that astronomers use to measure distance. The most certain way to determine distances unfortunately can be used only for nearby stars. This method employs the phenomenon of *parallax*, the apparent sideward displacement of a nearby object with respect to a background of more distant objects if we, the observers, move. Figure 4-12 shows how the motion of the earth around the sun produces a yearly back-and-forth displacement in the apparent position on the sky of a nearby star relative to more distant stars, which

71

appear essentially fixed. We see a similar parallax effect when we observe the sideward shift of a nearby object (such as our outstretched hand) with each eye alternately closed.

By knowing our average distance from the sun, called the astronomical unit (equal to 150 million kilometers), and by measuring the angular shift in position of the nearby star, we can use trigonometry to determine the distance to that star (see Figure 4-12). But how do we know the distance to the sun? The most accurate measurements of the earth-sun distance involve an even more fundamental method: use of the constant speed of light. By measuring the round-trip travel time of radar pulses beamed from earth and reflected back from the planets, the distances to the planets can be accurately determined. In addition, long and careful study of the motion of the planets has given us their distances in terms of the astronomical unit. Thus we can obtain the value of the astronomical unit by combining these two sets of observations: the round-trip travel time of the radar pulses and the observed motion of the planets across the sky. The value of the astronomical unit is now known to a precision of better than one part in 10 million.

If we try to use the parallax method to determine the distances to stars more than 100 light-years from us, we find that the stars' apparent motions are too small to be measured. Our atmosphere blurs our vision, making the measurement of angular displacements less than $\frac{1}{50}$ of a second of arc impractical (3,600 seconds of arc equal 1 degree). We must therefore use other, less accurate methods of distance determination. Unfortunately, all these other methods rely on *assumptions* about the intrinsic nature of the object whose distance we want to determine. We can neither board a spaceship and journey to the object in order to check the validity of our assumptions, nor can we obtain a parallactic "fix" on its distance from our motion. We must rely solely on the information brought to us by photons or other massless particles across tremendous reaches of space.

The Luminosity-Apparent Brightness Relation

A common method of estimating the distance to an astronomical object uses the relationship between the object's *luminosity* (energy radiated by it per unit time), an *intrinsic* property, and its *apparent brightness* (energy received by us per unit time per unit area of our detector), an *observed* property. Ignoring subtleties introduced by the laws of relativity, the observed brightness is proportional to the luminosity of the object and inversely proportional to the square of its distance. Thus an increase in the distance to a source of radiation by a factor of 10 reduces its apparent brightness by a factor of 100. This relationship can be understood by observing that the same amount of energy per unit time must flow through any imaginary spherical surface centered on a steady source of radiation.

Thus the amount of energy per unit time per unit *area* must decrease inversely with the area of the spherical surface, and therefore with the square of its radius (see Figure 4-13[a]).

Because we can measure the brightness of an object, we can find its distance if we know its luminosity. But how do we determine the luminosity? The best estimates rely on some observed property of the object that we believe is related to its luminosity. For example, for a nearby main-sequence star (whose distance is known from the parallax method), we find that the star's spectral characteristics (such as its surface temperature and the strengths of its spectral lines) are related to its luminosity. (We have already seen the temperature-luminosity relation in Figure 4-3.) If we assume that more distant stars resemble nearby stars, then we can determine a remote star's luminosity, and thus its distance, from the nature of its spectrum.

The Cosmic Distance Ladder

For distances at which main-sequence stars cannot be accurately studied by spectral analysis, we must find more luminous distance indicators. Bright variable stars called Cepheids have proven to be an extremely important indicator because their size and luminosity vary in a regular, cyclical manner. By studying nearby Cepheid variable stars, whose distances we know because Cepheids exist in clusters with other stars whose distances can be determined by other methods, astronomers have found a relation between a Cepheid's average luminosity and its observed period of brightness variation. The Cepheids can then be used to determine the distances to nearby galaxies, if we find stars in them that exhibit the familiar pattern of light variation and assume that the relation between the observed period of brightness variation and the average luminosity of a Cepheid variable persists from galaxy to galaxy.

This process is repeated each time a given type of distance indicator becomes too faint to be detected as a result of increasing distance. We must then find a more luminous type of source with an observed property that can be shown to be correlated with its luminosity. Then we can determine the greater distances at which we can see such a source, extending the "cosmic distance ladder" one more rung. (Objects of a certain type that are thought to have the same luminosity are called "standard candles." For instance, extremely bright stars, supernovae, and even certain types of galaxies have been used as such standard candles.) At each rung in the distance ladder, we encounter further uncertainty in using a particular type of luminosity indicator to make distance determinations. The total uncertainty thus becomes greater the farther we extend the method, since the accuracy of any such distance determination depends on the accuracy of all preceding rungs in the ladder.

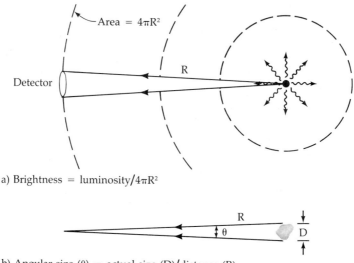

a) Brightness = luminosity/$4\pi R^2$

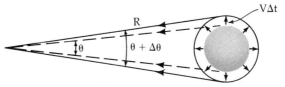

b) Angular size (θ) = actual size (D)/distance (R)

c) Change of angular size ($\Delta\theta$) = change of actual size ($2V\Delta t$)/distance (R)

Figure 4-13. This diagram shows three of the most common ways to determine the distance (R) from us (at the left) to remote objects. (a) We can use the relation between brightness (energy received per unit area of detector per unit time) and luminosity (energy radiated by the source per unit time). (b) We can use the relation between an object's observed (angular) size (θ) and its actual (linear) size (D). (c) For expanding, spherical objects, we can use the change with time of relation (b), which yields relation (c). $\Delta\theta$ denotes the change in angular size observed during a short interval of time (Δt). The object, expanding with velocity V, increases in actual size (D) by an amount $2V\Delta t$ during the same time interval Δt. The velocity V can be determined from Doppler shift measurements, if the expansion is spherical.

Angular Size as a Distance Indicator

A second method of determining distances suffers from the same problem—reliance on the indirect determination of an intrinsic property of the objects used as distance indicators. This alternate method employs the fact that the observed (angular) size of an object is proportional to its actual (linear) size and inversely proportional to its distance (see Figure 4-13[b]).

If we can find a relationship between the actual intrinsic size of the object and some observed property of the object, we can determine the object's distance from its observed angular size. We must find a new relationship each time the distance grows so great that the angular size of the indicator becomes too small to be accurately measured. Hot clouds of interstellar gas represent one such indicator that astronomers have tried to use. Thus far, distances derived from angular-size measurements are even more uncertain than those derived from brightness measurements, because the estimates of intrinsic sizes usually contain more uncertainty than the estimates of intrinsic luminosities.

The Motion Method

A third method of distance determination avoids many of the problems associated with the brightness and angular-size methods and dispenses with the need to construct a distance ladder. This method, which we shall call the motion method, can be applied most directly to spherically symmetric expanding (or contracting) objects. The observed rate of change of the angular size of such an object will be proportional to the object's expansion velocity and inversely proportional to its distance from us (see Figure 4-13[c]). We can determine the expansion velocity by observing the Doppler shift of the spectral lines produced at the surface of the object, for we assume that the object expands with the same speed in all directions. Hence if we can determine the rate of change of the object's angular size, we can find its distance. Astronomer Walter Baade first proposed this method in 1926 as a way of measuring the distance to variable stars.

There are, however, two problems with this method, both of which arise from the fact that the only objects for which this method has been used successfully are the exploding stars we call supernovae. Supernovae occur at such great distances that we cannot measure their angular sizes directly. How, then, can we know that the explosion of a particular supernova is spherically symmetric, and how can we determine its angular size? The answers come from study of the spectrum of the supernova, because the variation of its brightness with wavelength enables us to calculate the amount of radiation leaving the surface of the supernova, which in turn allows us to deduce its angular size and to estimate its sphericity.

This analysis, which requires the construction of a model of the physical conditions at the surface of the object, can be quite complex. Nevertheless, a supernova spectrum contains so much information that we should be able to deduce the required characteristics of the source. For a few weeks, a supernova is so luminous we can detect it at extremely great distances. (Afterward, it fades to obscurity over a period of several years.) This method of distance determination has not yet been widely used, but it appears promising. Astronomers will need some such reliable method to determine many key properties of the universe, in particular its rate of expansion and its deceleration.

Luminosity, Size, and Mass

Once we know the distance to an object, we can find its luminosity and size from the relations that we have already discussed. We determine the object's luminosity using the brightness formula, and its size using the angular-size formula (assuming that we can accurately measure the object's apparent brightness and angular size). The determination of the fourth basic property, the object's mass, requires the use of Newton's laws of motion and gravitation.

When we apply Newton's laws to any system of objects held together by their mutual gravitational attraction, a relationship emerges that we might call the *mass formula*. This states that the product of the size of the system and the average of the square of the velocity of the "particles" in the system will be proportional to the system's total mass. We can find the size of the system from its distance and its observed angular size, and the velocities from measured Doppler shifts in its spectrum. The particles in the system could be stars within a galaxy or galaxies within a cluster of galaxies.

Astronomy makes great use of this mass formula. By using a more refined version of the formula, astronomers can determine the masses of some stars that exist in binary systems. The formula also tells us immediately the average velocity of the nuclei within the sun (which is related to the sun's temperature), since we know the sun's mass and radius. But it is the mass formula's application to galaxies that has led to the most surprising results.

Invisible Matter

During the past two decades, radio astronomers have discovered that the hydrogen gas in the disks of spiral galaxies extends to distances beyond the stellar distribution we can see in photographs. Even more surprising was the discovery that the velocity of rotation of this outermost gas (obtained from the Doppler shift of the 21-centimeter spectral line emission produced by hydrogen atoms) exceeds what had been expected. Astronomers found that the rotational velocity of this gas remains constant at distances greater than the visible extent of a typical galaxy.

The mass formula relates the velocity at any specific distance from the center of some system to the total mass at all smaller distances from the center. The mass at larger distances does not contribute to the net gravitational force, which determines the velocity of the matter at that distance. The fact that the velocity of the hydrogen gas remains constant at larger distances means that the mass of the galaxy contained within a given distance from its center increases in proportion to that distance, even though no visible matter (i.e., stars) appears in the outer regions.

Astronomers must conclude that these galaxies contain a great amount of matter in some *invisible* form. The latest results show that this invisible mass is at least 10 times greater than the visible mass within galaxies. All forms of matter exert a gravitational force, but the light we see from galaxies tells us only how much mass exists in the form of ordinary (i.e., visible) stars.

In the same way that we apply the mass formula to the motion of the gas and stars within galaxies, we can use it to analyze the motion of entire galaxies within clusters of galaxies. Astronomers have found that an even greater fraction of the mass (about 99 percent) within some clusters of galaxies does not shine but instead exists in an invisible form.

What is this invisible matter? We do not know, but we can eliminate some possibilities. For instance, if it were ordinary gas like the interstellar medium, we would have detected the matter by either its absorption or its emission of electromagnetic radiation at some wavelengths. But the matter could reside in extremely low-mass stars (which have extremely low luminosities) or in black holes, or it could consist of neutrinos with nonzero mass, to name the three most popular current candidates. Because invisible matter appears to overwhelm the contribution of all visible forms to the total mass of the universe, the nature and extent of this matter obviously has enormous cosmological implications. We shall see in later chapters that the density of matter in the present universe determines its future: either eternal expansion or eventual collapse.

5

The Life of a Star

STARS PLAY A KEY ROLE in the universe: They liberate energy and synthesize most of the elements. To understand how they do this, we shall follow the evolution of a star as astronomers and physicists have deduced it: by using the laws of physics to construct mathematical models of stars, always checking to verify that the properties of these theoretical stars agree with the observed properties of real stars. With this approach, astronomers have come to understand much of a star's "life" in fairly complete detail. The exceptions lie at the boundaries: stellar birth and stellar death.

Star Formation

The birth of a star is perhaps the least understood phase of its life. We observe young stars in galaxies today, immersed in clouds of gas and dust. We may naturally conclude that these stars have condensed from the material that forms these clouds. We are familiar with the way water vapor condenses to form droplets in our atmosphere. For a star, however, gravitation, not the electromagnetic force of chemical binding, drives the condensation process, which occurs on a scale 10^{18} times larger than the size of a raindrop. Stars apparently have formed from the gravitational collapse of clumps of gas—regions of somewhat higher-than-average density within an interstellar gas cloud.

A typical cloud contains enough material to make thousands of stars. But how do these clumps arise? Various theories exist, but the process remains imperfectly understood. The same difficulty arises in explaining how stars formed in the distant past and indeed, as we shall see, in explaining the existence of galaxies themselves.

Once formed, a clump of gas tends to collapse under the attractive force of its self-gravitation. The major force opposing the collapse arises from the distribution of pressure within the collapsing clump, called a protostar.

The pressure force pushes outward because the density and temperature of the gas (whose product determines the gas pressure) decrease outward from the center of the protostar. The competition between these two forces, gravitation and pressure, controls the structure of the star throughout the remainder of its life.

The pressure forces eventually halt the collapse of the star, but only after it has grown hotter than the cloud from which it was born. The mass formula tells us that the smaller the star's radius, the higher its internal temperature (proportional to the square of the particle velocities). Hence collapse produces an increase in temperature within the protostar. Since the star is hot, it radiates energy, at first primarily in the infrared portion of the spectrum. But the laws of motion and of gravitation also tell us that as a star loses energy, it must slowly contract, and therefore must continue to increase its temperature. A star behaves in this sense quite differently from objects familiar to us: As a star loses energy, it heats up!

The Onset of Nuclear Fusion

The slow collapse of the star continues for about a million years, until the star's central temperature reaches 10 million degrees Kelvin, a thousand times hotter than its surface. At this time the protostar becomes a star because nuclear fusion begins in its interior, supplying enough energy to match that lost by radiation (starlight) from its surface. Henceforth, nuclear fusion, rather than collapse, furnishes the star's radiant energy. Nuclear fusion requires an enormous temperature because the resulting high velocities of the positively charged nuclei are needed to allow some of them to overcome their mutual repulsion and approach one another closely enough to fuse. (At these temperatures, most atoms have been stripped of all their electrons, so we need consider only the bare nuclei.) Figure 5-1 shows some of the nuclear reactions that occur deep within stars like the sun. Only light nuclei undergo fusion, because they have the smallest electric charges, and therefore the least electric repulsion. Heavier nuclei can fuse only at higher temperatures.

The first step in the chain of fusion reactions is the merging of two protons to form a deuteron (an isotope of hydrogen that contains a proton and a neutron), a positron (the antiparticle of the electron), and a neutrino. This fusion is so improbable that a typical proton undergoes such a reaction once every 10 billion years or so. However, because there are about 10^{57} protons in the sun, 10^{39} of these reactions occur every second.

Each fusion produces an amount of additional kinetic energy that is proportional to the difference in mass between the particles before and after the reaction. This process represents the conversion of mass into energy, first predicted by Einstein in his equation $E = mc^2$ (energy equals mass times the square of the speed of light). The entire cycle of fusion

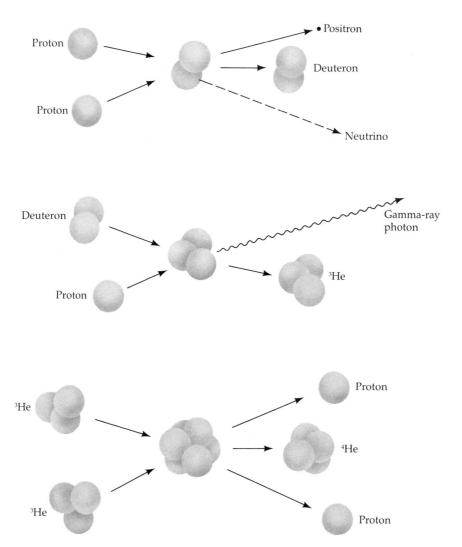

Figure 5-1. In most stars the proton-proton cycle converts hydrogen to helium chiefly through three steps of nuclear fusion. In the first step, two protons interact to produce a deuteron (a bound proton and neutron), as well as a neutrino and a positron. In the second step, the collision of a deuteron with a proton produces a ^{3}He nucleus (containing two protons and a neutron) and a photon. Finally, two ^{3}He nuclei collide to produce a ^{4}He nucleus (containing two protons and two neutrons) and two protons. In each step, energy is released because the combined mass of the outgoing particles (right-hand side) is less than that of the incoming particles (left-hand side). The difference in mass (Δm) is converted into an amount of energy (ΔE) given by Einstein's formula $\Delta E = \Delta mc^2$.

reactions, called the "proton-proton cycle," results in the conversion of four protons into a helium nucleus (^{4}He) plus photons and neutrinos. (The positrons quickly annihilate.) Although the proton-proton cycle converts only about 1 percent of the mass into energy, it nevertheless produces a considerable amount of energy. If, for example, you could convert 1 percent of the mass of your body into energy, you could supply the electric power requirements of 10 million homes for a year.

Looking into a Star

Each star is a ball of hot gas, in which the temperature and density decrease outward from the center. Because our sun, a typical star, has a radius of about 2 light-seconds, the neutrinos produced in fusion reactions at the sun's center emerge at the surface about two seconds later. Most neutrinos do not interact at all with the sun's matter as they pass through it. By contrast, the photons produced at the center interact readily, undergoing many scatterings, absorptions, and reemissions on their way to the surface. The photons' paths therefore are not straight, but in fact are so convoluted that a typical photon takes 10 million years to reach the sun's surface. If the nuclear reactions at the center of the sun ceased today, we would not know this for 10 million years were we able to detect only the sun's photons.

However, it now appears that neutrinos from the sun's center have been detected. A remarkable experiment under the guidance of the chemist Raymond Davis of the Brookhaven National Laboratory detects the conversion of a single atom of chlorine to one of argon, which occurs upon the absorption of a neutrino by the atomic nucleus. (Actually, the neutrinos Davis can detect are produced in the sun by a less common cycle of reactions than that shown in Figure 5-1, one involving the fusion of ^{3}He nuclei with ^{4}He nuclei.) The chlorine resides in a tank containing 400,000 liters of cleaning fluid, buried deep within a gold mine in South Dakota to shield the fluid from the competing effects of cosmic rays. Although the number of neutrinos that have been detected falls somewhat below the number predicted by most theoretical models of the sun, these neutrinos show that we do understand the basic process by which the sun and other stars produce the energy they radiate.

Stellar Lifetimes and Masses

How long can a star remain on the main sequence, converting hydrogen into helium? A star's lifetime in any phase of its existence must be proportional to the energy available from the particular nuclear reactions that can occur during that phase. This available energy is proportional to the number of nuclei, and therefore to the mass of the star. The lifetime also must vary inversely to the rate at which the star loses energy, that is, to its luminosity. The luminosity of more massive stars far exceeds that of less massive stars, varying approximately as the cube of a star's mass. Hence the most massive, most luminous stars have the shortest lifetimes, even though they have the greatest supply of energy. The most luminous stars yet observed will consume their hydrogen fuel in about 1 million years, whereas the sun will take a little over 10 billion years.

Why do all stars have masses not greatly different from that of the sun? Calculations show that balls of gas with less than about $\frac{1}{10}$ of the sun's mass cannot produce (through their self-gravitation) a central temperature hot enough to ignite nuclear reactions. Such objects must resemble giant Jupiter-like planets. The paucity of extremely massive stars has a slightly more complicated explanation. Calculations show that the pressure within such massive stars is produced primarily by photons, rather than by electrons and nuclei. This circumstance tends to make the stars unstable. Theoretical models predict that such massive stars either expel a large fraction of their mass or else collapse because the pressure force cannot balance the gravitational force. In fact, stars with more than 50 times the sun's mass are exceedingly rare.

The Red-Giant Phase

What happens to a star when it exhausts its supply of hydrogen? The star must again contract, and, as we have previously discussed, this contraction will raise the star's central temperature. The resulting higher temperatures permit the heavier nuclei (which have greater electric repulsion because their charges are greater) to interact. Fusion reactions among heavier nuclei represent a fresh source of energy that allows the star to shine steadily again until the reacting nuclei become exhausted in their turn.

The "second wind" for a star is produced by the "triple-alpha process," the conversion of three helium nuclei into a carbon nucleus. This helium fusion requires a central temperature a few times higher than that during the star's main-sequence phase. In this second phase, the star becomes a "red giant." Although its inner regions contract, its outer envelope swells enormously in size. The star contains a bloated, rarefied envelope surrounding a dense energy-producing core. As the star's central temperature increases, its surface temperature falls because of its increasing size. Hence it appears redder. Such stars thus evolve toward the cooler (red) portion of the giant region of the temperature-luminosity diagram (see Figure 4-3). Helium fusion produces less energy than hydrogen (proton) fusion, but the star's luminosity increases as it becomes a red giant. The time it takes a star to evolve through this red-giant phase must therefore be much shorter than the time it spends on the main sequence.

These changes in the properties of a star as it enters its red-giant phase allow astronomers to measure the ages of clusters of stars. It is natural to assume that the stars within a cluster all formed at approximately the same time. However, the stars are born with a wide range of masses. Consider how a plot of luminosities and temperatures of stars (such as Figure 4-3) would appear for stars in a given cluster. More luminous (and massive) stars spend less time as main-sequence stars before evolving into red giants. Hence as the stars in a cluster age, first the most luminous and then the slightly less luminous stars end their main-sequence phases, evolving to the red-giant region of the temperature-luminosity diagram. The luminosity (or the surface temperature) of the most luminous stars remaining on the main sequence reveals the age of that cluster of stars. (Figure 4-3 shows the ages at which stars leave the main sequence.) Using a more refined version of this technique, astronomers have derived an age of approximately 12 to 18 billion years for the oldest star clusters found in our galaxy.

For any star, the process of contraction, accompanied by an increase in central temperature, starts anew each time the lightest available nuclei are completely transformed into heavier nuclei at the star's center. Many stars seem "discontent" in the physical state they reach following their red-giant phase. Some stars then eject their outer layers into the interstellar medium, forming the beautiful shells of gas we call planetary nebulae (see Figure 5-2). Other stars pulsate, producing the variations in brightness that characterize the Cepheids and other such variable stars.

Deep inside the star, nuclear fusion reactions not only produce energy but also create new types of nuclei. As a result, the star's composition changes with time, and the composition at any given time varies with distance from the center. The outer layers of the star have not reached temperatures high enough to ignite reactions among the heavier nuclei. Hence they retain more of the star's original composition. The closer we come to the center of the star, the heavier are the nuclei we find, for the gas has reached higher temperatures and has therefore undergone more types of fusion reactions.

The Death of a Star

Evolution of a star beyond the red-giant phase depends critically on the star's mass, the basic factor that determines its final fate—whether it will become a white dwarf, a neutron star, or a black hole. Before we can understand this final phase of stellar evolution, we must introduce a new physical concept: degeneracy.

Particles in a gas at extremely high density obey the laws of quantum mechanics rather than the familiar equations of classical mechanics. Because of the Exclusion Principle (see Chapter 3), we find that as we add more and more fermions to any region of space, the particles must occupy states of higher and higher energy since these are the only states available.

Figure 5-2. This planetary nebula in the constellation Aquarius shows its central star only dimly in a visible-light photograph, because the star now emits most of its energy as ultraviolet photons. The shell of gas expelled from the aging star glows because these ultraviolet photons excite electrons into higher-energy orbits in atoms within the gas. As the electrons cascade into lower-energy orbits, the atoms emit visible-light photons. (Photo: Palomar Observatory.)

Physicists call this forcing of additional fermions into states of progressively higher energy a condition of "degeneracy."

The pressure that a gas exerts depends upon the kinetic energies of its constituent particles. The pressure in an ordinary gas is proportional to its temperature (and its density), for temperature is a measure of particle energies. The pressure of a gas at sufficiently high densities, however, can be appreciable even at zero temperature, because the particles within the gas acquire their kinetic energies through the quantum-mechanical effect just described. This new type of pressure is called "degeneracy pressure." Armed with this information, we can investigate the final stages in the life of a star.

White Dwarf Stars

In stars with masses up to about 1.2 times the mass of the sun, the electrons become degenerate soon after the star's red-giant phase. (The original mass of the star could have been greater than this critical value if the star lost mass as it evolved.) With its electrons providing degeneracy pressure, the star can exist forever in this state of balance. The support from degeneracy pressure persists even as the star cools. The radius of such a star roughly equals that of the earth, which makes it a "dwarf" star. This small size implies that the density of the matter within the star must be enormous—about a million times greater than that of ordinary solids and liquids, although the matter remains gaseous. One teaspoon of such matter would weigh about a ton at the earth's surface.

The star, once the core of a red giant, has a surface hot enough to radiate visible light and hence has the name "white dwarf." Since it now produces no energy in its interior, it cools slowly, eventually becoming a "black dwarf." White dwarfs have lower luminosities than most main-sequence stars but similar surface temperatures (see Figure 4-3).

In 5 to 8 billion years the sun will become a red giant, an event likely to destroy any life on earth. After a billion years or so as a red giant, the sun will probably shed its outer layers and enter peaceful, eternal retirement as a white dwarf fading into blackness.

Supernova Explosions

Stars too massive to become white dwarfs—because their interiors do not develop sufficient electron degeneracy pressure to provide ongoing support of the additional mass—continue to fuse heavier and heavier nuclei until they produce a core of iron. At that point nuclear fusion reactions can release no more energy, so the star collapses. This collapse raises the star's density tremendously. When the density of matter at the star's center reaches the enormous value found within nuclei, two significant events occur. First, the composition of the core of the star changes from a gas of primarily iron nuclei to a gas of primarily free neutrons. In addition, the neutrons become degenerate, and their degeneracy pressure halts the collapse of the core. The core bounces back to a slightly larger size, like a tightly squeezed rubber ball, generating a shock wave that may propagate outward, ejecting the outer layers of the star into space. However, theoretical models of supernovae still have difficulty reproducing this expected behavior.

Figure 5-3. The Crab Nebula is the remnant of a supernova explosion that was observed in 1054. The explosion violently ejected matter and left behind one of the stars near the center of the nebula. This star is in reality a pulsar, a rapidly rotating neutron star whose beam of electromagnetic radiation sweeps over us 30 times per second. The nebula glows because high-energy photons and other particles excite atoms in the gas, and because high-energy charged particles produce "synchrotron radiation" as they spiral in magnetic fields created by the supernova. (Photo: Lick Observatory.)

We know that some stars eject their outer layers in this way because we see the matter dispersed throughout the surrounding interstellar medium (see Figure 5-3). Astronomers call such an explosive event a supernova to distinguish it from a milder form of explosion known as a nova. For a few weeks, the expanding shell of gas glows as brightly as a billion stars, producing a spectacular event such as the one Tycho saw in 1572. Most importantly for us, the explosion ejects the heavy elements previously synthesized within the star, further enriching the composition of the interstellar medium.

Neutron Stars and Pulsars

Physicists believe that what remains at the center of the supernova explosion usually (if not always) forms a neutron star. The matter within neutron stars has roughly the same density as an atomic nucleus and may be pictured as a giant nucleus made of neutrons plus a small percentage of protons and electrons. A neutron star's radius is only a few kilometers, yet its mass is about twice that of the sun (hence almost a million times that of the earth).

Neutron stars were only a theoretical prediction until 1967, when Anthony Hewish, an astronomer at Cambridge University, and his student Jocelyn Bell discovered strange objects now known as "pulsars." What they detected were bursts of radio waves, arriving at regular intervals from certain directions on the sky. Although they considered that the signals might be coming from other civilizations somewhere in our galaxy, they quickly ruled out this possibility. The true origin of the pulses is only slightly less exciting: rotating, magnetized neutron stars. As has frequently occurred in astronomy, no one predicted the phenomenon of pulsars, although the underlying object, a neutron star, had been studied theoretically for 30 years. This discovery provides another example of the way in which a careful observational survey of the sky may reveal something totally unexpected.

Soon after the discovery of pulsars, astrophysicists developed a pulsar model based upon a neutron star with an intense magnetic field. Although stars possess magnetic fields, the neutron star's field must be 10^{12} times stronger than that of an ordinary star like the sun. This is not surprising, because a star's magnetic field increases in strength as the star contracts, and a neutron star is 100,000 times smaller than an ordinary star. A star also tends to spin more rapidly as it collapses, so we expect neutron stars to be rotating rapidly.

The rotation of the neutron star plays a crucial role in the mechanism that generates the observed pulsar radiation. It produces an electric field that accelerates particles near the star. These particles in turn radiate electromagnetic waves when they reach speeds close to the speed of light. The magnetic field guides the particles, producing "searchlight" beams of radiation that rotate with the pulsar. Each time a beam sweeps by the earth, we observe a pulse of radiation.

This model received brilliant confirmation in 1969 with the discovery that a "star" at the center of the Crab Nebula supernova remnant (see Figure 5-3) is actually a pulsar emitting a broad spectrum of electromagnetic pulses 30 times per second. Because this interval is shorter than our

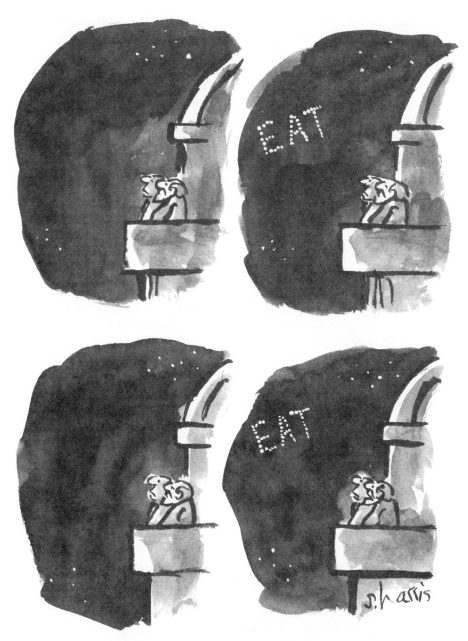

"I don't care what it looks like—they're pulsars."

eyes can detect, the "star" appears to shine steadily. We rely on special electronic detectors to reveal its rapid light variations.

Astronomers also found that this neutron star (like all other pulsars) steadily loses energy: Its rotational velocity slowly decreases. This loss of rotational energy just matches the energy needed to produce the emission observed from the supernova remnant. To astrophysicists, this clinched the argument for the model of a pulsar based upon a neutron star. A pulsar represents the most efficient particle accelerator we know—it converts rotational energy into high-energy particles with an efficiency close to 100 percent.

Many neutron stars may not produce detectable pulsars because they rotate too slowly to emit enough radiation for us to detect them. In addition, it is not clear that all supernovae leave behind a neutron star or that the collapse into a neutron star necessarily leads to a supernova. What we do know is that neutron stars exist, millions of them in our galaxy (and by inference, in other galaxies).

Black Holes

The final possible fate of a star, and the most dramatic of its destinies, is to become a "black hole." Stated simply, a black hole is a region of space from which nothing, not even light, can escape.

To obtain some idea of how such a strange situation can occur, we may consider a rocket launched from the earth's surface that quickly acquires a velocity of at least 11 kilometers per second (the earth's escape velocity) before its engines are shut off. Then the earth's gravitational force cannot ever make the rocket reverse its direction of motion and fall back to earth. Now suppose that we could make the earth smaller and smaller without changing its mass. This would make the earth's gravitational force at its surface stronger and stronger. Eventually the earth's gravitational pull on the rocket would be so strong that the rocket's escape velocity would have to equal the speed of light. If the earth were any smaller than its size at this point, the escape velocity would have to exceed the speed of light. Because nothing can in fact surpass the velocity of light, nothing would be able to escape the pull of the earth's gravitational field, and the earth itself would have to collapse. Even photons could not escape. Yet particles could still fall into this imaginary collapsed earth, which has thus produced a black hole around itself.

The theory of general relativity predicts that such a region of space, invisible to us, has a radius that is proportional to the mass of whatever collapsed to produce it. This critical radius, or "gravitational radius," equals 3 kilometers times the mass measured in terms of the sun's mass. We know that the degeneracy pressure of electrons or neutrons will eventually halt the collapse of any star with a mass less than or equal to about twice that

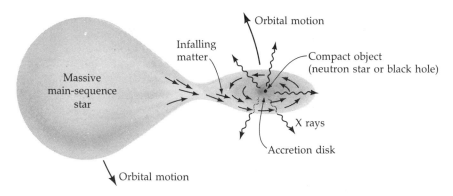

Figure 5-4. Compact objects (neutron stars or in some cases possibly black holes) have been found to exist in some binary star systems. In such a system, matter from the ordinary star slowly spirals toward the compact object, forming a rotating accretion disk. Near the compact object, the disk becomes hot enough to emit X rays.

of the sun. But a star that retains more mass than this cannot be prevented by any form of pressure from eventually collapsing completely. The star then disappears from view as a black hole forms around it. Yet the gravitational force exerted on distant bodies remains the same as that produced by the original star because the mass of the collapsing object itself cannot change.

Although astrophysicists expect evolved massive stars to collapse into black holes, a massive star that expels much of its mass in the process of evolution, or rotates so rapidly that it fissions into two or more stars, can avoid this fate. Nonetheless, most astrophysicists believe that a significant number of black holes exist throughout the universe.

The best black-hole candidates are objects in mutual orbit with an ordinary star (see Figure 5-4). The most likely of these binary systems to contain a black hole is named Cygnus X-1: It emits X rays and lies in the direction of the constellation Cygnus. Cygnus X-1 belongs to a class of binary X-ray sources, discovered only during the past decade by detectors in orbit above the earth's atmosphere. By analyzing the spectrum of the visible star in the pair, astronomers can determine the star's orbital velocity (which produces a variable Doppler shift) and its mass. From Newton's laws of motion and gravitation, it is then possible to determine what the mass of the unseen companion must be. The dark companion in Cygnus X-1 has a mass about ten times that of the sun.

We also know that the unseen companion must have an extremely strong gravitational field near its surface, and hence must be tiny though massive. This conclusion is based on the following evidence. Gas (stripped from the companion star) slowly but continuously spirals toward the unseen object in the form of a disk. The gas heats up significantly as it spins faster and faster while approaching the object, reaching a temperature sufficient to produce the X-ray photons we observe. This heating through gravitational attraction requires a strong gravitational field. We know that such a field could not arise from a neutron star (the only other object that can produce such a strong gravitational force) because a neutron star's mass cannot exceed twice the sun's mass. Hence the only remaining possibility that appears reasonable is that the Cygnus X-1 system contains a black hole, although we cannot yet consider this evidence to be conclusive. But today most astronomers find Cygnus X-1 fairly persuasive evidence that a black hole with ten times the sun's mass exists in our galaxy, some 10,000 light-years from earth.

Supermassive Black Holes

Astronomers have also found some evidence for the existence of black holes at the centers of the violent galaxies we examined in Chapter 4. In such exploding galaxies the tremendous energy of the ejected matter and the high luminosity of the source of the explosion require that the "engine" powering the activity be immensely massive—at least 10 million times more massive than the sun. Yet we also determined that this engine cannot be particularly large. The presently favored theory posits that an extremely massive black hole lies at the heart of the engine.

As with less massive black holes, the source of energy could be the infall of material from an accretion disk around it (see Figure 5-4). The gas would not come from a single star, but would instead be collected from the interstellar medium and stars in the region around the massive black hole. The jets of material that we saw in Chapter 4 are presumably directed outward along the axis of rotation of the black hole. Although many detailed models have been proposed, we do not yet have a clear idea of the mechanism by which the energy of the matter spiraling in toward the black hole is converted into the energy in the jets. We do know that such a conversion must occur before the matter has fallen within the critical gravitational radius, where it disappears forever, increasing the mass of the black hole.

More direct evidence for the presence of some sort of massive, compact object at the centers of galaxies has come from the distribution and velocities of stars in the central region of the elliptical galaxy M87 (see Figure 4-9). The stars' motions apparently reflect the controlling gravitational force of a tremendous amount of matter (at least a billion solar masses), contained within a volume smaller than can be seen directly (about 100 light-years across).

A quite different way to discover black holes may prove feasible in the future. This method relies on detecting the burst of gravitational radiation

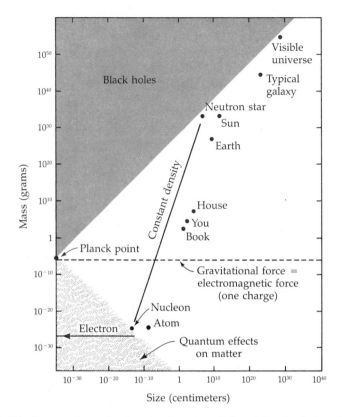

Figure 5-5. This diagram shows the domain of our knowledge of the physical world; objects are classified by their masses and sizes. Objects in the black-hole region cannot be seen. Objects in the quantum region (lower left) do not obey the familiar equations of classical mechanics. The electron is indicated by a line, rather than by a point, because elementary particles do not have measurable sizes. Any two objects, each of which has a single unit of electric charge and a mass of 2×10^{-6} grams (lying along the central, dashed horizontal line), produce equal electromagnetic and gravitational forces on each other, regardless of their separation. Samples of matter with various sizes and masses but the same particular density fall along a line parallel to that marked "constant density." The line drawn here shows that neutron stars and nucleons (or nuclei) have the same density but very different sizes and masses. In addition, notice that an atom, a book, you, the earth, and the sun all have about the same density even though the sizes and masses differ greatly. The black-hole and quantum boundaries meet at the Planck point (mass = 2×10^{-5} grams, size = 2×10^{-33} centimeters), where the effects of strong gravitational fields (black holes) and quantum mechanics are both important. Distances less than the Planck size do not appear in the diagram because we do not yet know the laws of physics that govern that domain.

generated by the collapse of the object that produced the black hole. If a collapsing star deviates significantly from spherical symmetry (for instance, because of its rotation), then the motion of the infalling matter and the rapidly changing strong gravitational field associated with the formation of a black hole should produce a short, intense burst of gravitational radiation. Gravitational-wave detectors already in existence could detect such an event if it were to occur within our galaxy. However, knowing the rate at which stars die, we expect no more than one such event every few years within a galaxy such as our own.

The Domain of Our Knowledge

Let us summarize our inventory of the universe, carried out in Chapters 3, 4, and 5, by illustrating in Figure 5-5 the enormous range in the scales of the structures that we have found. Notice that there are two boundaries in this diagram. One separates all visible objects from those that have collapsed to form black holes. The other boundary separates the region where the laws of quantum mechanics, rather than those of classical mechanics, govern the behavior of objects. The two boundaries meet at the "Planck mass" (10^{-5} grams) and the "Planck distance" (10^{-33} centimeters), named after the German physicist Max Planck. At this point on the diagram, the effects of strong gravitational fields (black holes) and quantum mechanics merge. When dealing with such conditions, we must replace the classical theory of gravitation—general relativity—with some quantum theory of gravitation that is still undeveloped. (We shall have more to say about this in Chapter 8.)

Near the black-hole boundary, we find most astronomical objects, since their structure is controlled by gravitation. Near the quantum domain, we find atomic systems (those containing a small number of particles). In between, we find those objects with which we are most familiar in our daily lives.

Armed with this knowledge, we can now venture to realms more distant than our cosmic neighborhood.

6

Messages from the Farthest Reaches

THE GREAT EPOCH of modern cosmological discovery dawned early in this century with the construction of large telescopes, in particular the 2.5-meter (100-inch) reflector at the Mount Wilson Observatory near Pasadena, California, in 1917 (see Figure 6-1). The increased light-collecting area of the mirrors in these reflecting telescopes allowed astronomers to identify individual stars in the nearest galaxies, which in turn led to an understanding of the nature and distances of these galaxies. These large telescopes also provided our first views of the deeper regions of space, which Edwin Hubble, the first person to chart these regions, called "the realm of the nebulae [galaxies]."

As telescopes have increased in size and sensitivity, astronomers have probed farther and farther reaches of space. And as we have extended our "vision" to portions of the electromagnetic spectrum other than that of visible light, the unexpected has often emerged in the flood of information carried by these ancient photons.

Three major discoveries have revealed how the universe behaves at the largest scales of distance: (1) the distribution of galaxies, (2) the expansion of the universe, and (3) the microwave background radiation. In this chapter we shall investigate what these discoveries tell us about the present universe. In the next chapter we shall engage in the more speculative investigation of what they may tell us about the past and future of the universe.

Before proceeding, let us anticipate one important aspect of the universe that will emerge from the model we shall construct. We shall see in the next chapter that photons have been able to travel freely through the universe only since a certain definite time in the past. Since the speed of light (or of any photon) is finite, any electromagnetic radiation we detect has come to us from only a limited portion of the universe. (The size of this

Figure 6-1. The 100-inch (2.5-meter) reflecting telescope at the Mount Wilson Observatory in California has revealed many of the important features of our universe. (Photo: Mount Wilson Observatory.)

"visible universe" does increase with time, but the increase is significant only over billions of years.) Astronomers believe that we have already reached this limit of detectability at microwave wavelengths and are now proposing to build telescopes capable of approaching this limit at other wavelengths. Hence the present generation of astronomers has a unique opportunity: to be the first to observe most of the matter that we shall ever be able to observe.

We saw in Chapter 2 how astronomers estimated the distances of certain "nebulae," proving that they are galaxies well beyond our own. Following this realization, in the decades of the 1920s and 1930s Edwin Hubble and his colleagues made two other major discoveries. One was the nature of the distribution of galaxies; the other was the expansion of the universe.

The Distribution of Galaxies

With the large telescope at the Mount Wilson Observatory, Hubble and other astronomers detected the light from galaxies as distant as a few hundred million light-years. They photographed the sky in many directions, then counted the number of galaxies visible on each photograph. Later astronomers have confirmed and extended their findings. Figure 6-2 is a computer-generated map, a composite of data compiled over 12 years, of the distribution of galaxies visible from the Lick Observatory near San Jose, California. The map includes a million galaxies, the most distant of which is a billion light-years away. Here we view the texture of the universe, that is, the distribution of visible matter on large scales of distance. What is the nature of this distribution?

To interpret a map such as Figure 6-2, we must allow for the fact that the earth blocks our view in certain directions (below the southern horizon when we observe from northern latitudes). In addition, interstellar dust in our galaxy absorbs the light from distant objects, obscuring our view of deep space in directions through the plane of the Milky Way. A still greater difficulty in interpreting our observations of galaxies is that we do not accurately know the distances to most of them. Nevertheless, allowing for some margin of error, we can obtain from such maps an intriguing picture of the distribution of galaxies in space.

On the smallest intergalactic scales of distance, many galaxies belong to clusters. These clusters range from loose groupings of a few dozen galaxies to the rich clusters (such as the Hercules cluster in Figure 4-8) that contain up to a few thousand galaxies within a diameter of roughly 20 million light-years. As we have seen (in Chapter 4), the mass necessary to hold the clusters together is much greater than that visible within its member galaxies. X-ray telescopes orbiting the earth have detected radiation from these rich clusters. This radiation can come only from a hot gas that exists between the galaxies. However, the mass of this gas falls considerably short of that needed to hold such a cluster together by gravitation.

When we consider larger scales of distance (corresponding to larger areas of the sky in Figure 6-2), we find that the clustering of galaxies becomes less pronounced. Think of viewing the sky in different directions through an aperture of a given size, corresponding to a certain amount of area on the sky. As we increase the size of the aperture, the number of galaxies we see in any direction will of course increase, but the relative number of galaxies seen in different directions through the same aperture will tend to become equal. As Hubble's work first showed, the distribution of galaxies looks the same in all directions on the sky, so long as we consider sufficiently large areas of the sky. We call the property of having the same appearance in all directions *isotropy*. This large-scale isotropy also describes the distribution of the more distant galaxies and quasars, some of which are visible only at radio wavelengths.

Figure 6-2 shows how galaxies are distributed two-dimensionally over the sky. We have two ways to determine how galaxies are distributed

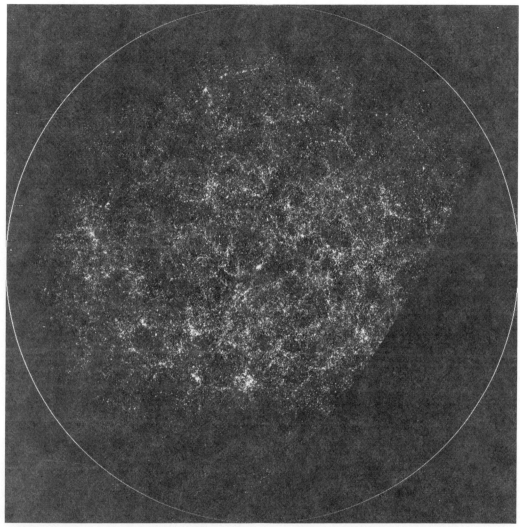

Figure 6-2. This map shows the distribution of galaxies over the sky visible from Lick Observatory. The center corresponds to a viewing direction perpendicular to the plane of our galaxy; the absence of galaxies around the edge is caused by the absorption of light in directions through the plane of our galaxy. (The sharp edge indicates the southern horizon.) The degree of brightness of each small dot represents the number of galaxies in the corresponding area ($\frac{1}{6}°$ by $\frac{1}{6}°$). The brightest dots represent ten or more galaxies. A million galaxies brighter than a certain fixed value are included in this map, at distances up to a billion light-years. (Photo: *CoEvolution Quarterly*. Original negative by M. Seldner, B. Siebers, E.S. Groth, and P.J.E. Peebles.)

through three-dimensional space on the largest distance scales, which are of the greatest interest to cosmologists.

The first approach invokes the Copernican Principle: Our position in the universe is typical, not special. It follows that an observer in any other galaxy would also see galaxies distributed more or less evenly over the sky. Mathematicians have shown that a universe that appears isotropic to all observers must be uniform—*homogeneous*—as well. (A property that is homogeneous is the same at all points in space; one that is isotropic is the same in all directions from a point in space.) Therefore, the properties of the universe, when averaged over sufficiently large volumes, must be the same *everywhere* at any given time, though they can change with time (but in the same way for all observers).

The second, and more reliable, approach to the determination of how galaxies are distributed through space is direct observation. In one method, astronomers determine how many galaxies are brighter than various arbitrary minimum values. From the brightness law (observed brightness is proportional to intrinsic luminosity divided by distance squared), we can calculate that the number of visible galaxies should be inversely proportional to the 3/2 power (the cube of the square root) of this minimum brightness, if galaxies are distributed uniformly throughout space. (Here we are neglecting effects of relativity and evolution.) For instance, if we double the diameter of the telescope mirror we use (and thereby increase its light-collecting area by a factor of four), we collect four times as many photons and thus see galaxies four times fainter than we could see before. We can therefore see galaxies twice as far away as before (by the brightness law) and accordingly should see eight times as many galaxies, since we observe a volume eight times greater (the volume we see is proportional to the distance cubed). Note that 8 is the 3/2 power of 4, in accordance with the formula stated above. Hubble and his successors indeed found this relationship to hold true for galaxies at distances greater than about 100 million light-years. This second approach leads to the same conclusion that follows from the first: The universe, so far as we can tell, appears to be homogeneous as well as isotropic on the largest distance scales.

The Texture of Universal Matter

We have seen what the universe looks like when we scan the sky in various directions. Let us now convert this two-dimensional picture into a three-dimensional picture of what the universe looks like on various distance scales by considering samples of the universe contained within imaginary cosmic cubes of increasing size.

If we imagine cubes with edges 10 million light-years in length, then some cubes would contain a few galaxies, some would contain one, and some would contain none. A very few would contain most of a rich cluster of galaxies. (We are discussing here the most luminous galaxies. A larger number of fainter galaxies do exist, but they contribute negligibly to the total mass and the total luminosity of the galaxies within our cube.)

When we increase the length of the edges of our cosmic cube 10 times, to 100 million light-years, each cube contains thousands of galaxies. Here the differences between cubes are not so dramatic as before, but some cubes do contain significantly more galaxies than the average. These giant agglomerations of galaxies are clusters of clusters, or superclusters of galaxies. Other cubes contain significantly fewer galaxies than the average, implying the existence of voids in the distribution of galaxies.

When we examine cubes with edges greater than 100 million light-years, the different samples of the universe all begin to look the same. On this scale, the number of galaxies within a cube divided by the volume of the cube (which is the average density of galaxies in that region of space) reaches a fixed value, the same for all cubes larger than a few hundred million light-years. Hence the total mass of galaxies per unit volume also approaches a fixed value—the present galactic mass density. This density plays a crucial role in the dynamics of the universe, but we do not yet know its value accurately. If we imagine all the matter within galaxies to be dispersed uniformly throughout space, we would have a gas containing less than one gas atom (plus the more massive contribution of invisible matter) per cubic meter. This density of atoms is 10^{27} times less than the density of our atmosphere.

What evidence do we have that matter exists in the apparent emptiness between galaxies? Recently, astronomers have found distant clouds of gas by the absorption lines they produce in the spectra of quasars. And some of the X-ray radiation that fills the sky could be produced by hot intergalactic gas, though other explanations are possible. In any event, we have no evidence that the mass of any such intergalactic gas exceeds even the visible mass contained within galaxies. Therefore its contribution to the total mass of the universe is probably negligible. We shall see, however, that a sea of photons fills the universe. In addition, undetected forms of matter may exist between the galaxies.

The Cosmological Principle

The conclusion that the universe presents the same large-scale aspect in all directions (isotropy) and at all points (homogeneity) has far-reaching consequences. This proposition has generally been elevated to the status of a first principle, called the Cosmological Principle. Of course, that does not prove its necessary truth since, like any other theoretical proposal, it remains subject to observational refutation. Until the Cosmological Principle fails to agree with observation, we accept it as valid because it explains (as do the laws of physics) in the simplest way what we observe.

The Cosmological Principle immediately rules out the possibility that the universe has either a center or an edge. For instance, if the universe resembled a giant cluster of galaxies, then it would appear different in different directions, unless we happened to be at its center (which would violate the Copernican Principle). In addition, the density of galaxies would likely be lower near the edge than near the center of the cluster. We can, however, never rule out the possibility that the size of such a giant cluster far exceeds the size of the visible universe, in which case the observable properties of the cluster would be essentially the same as those of the uniform universe posited by the Cosmological Principle.

The Geometry of the Universe

The Cosmological Principle also allows us only a few possibilities for the overall geometry of the universe. We may ask: Why isn't there only one possible geometry, the Euclidean geometry that underlies all our measurements of distances, areas, and volumes? The answer is gravity. Because gravitation affects all forms of matter, it affects all of our measurements. The results of these measurements, which we can describe in terms of some particular geometry, are therefore influenced by the gravitational field that exists. The Cosmological Principle restricts the distribution of matter, which is the source of gravity, and hence restricts the structure of the large-scale gravitational field throughout the universe. Therefore, it restricts the geometry of the universe. If we were to abandon the Cosmological Principle, the possibilities for the large-scale, three-dimensional geometry of the universe—the structure of space—would grow immensely.

Euclidean geometry works well for us on earth because the earth's gravitational field is relatively weak, and we measure relatively short distances. But the gravitational field of the universe can produce important effects over the large distances involved in cosmology. Since we cannot lay yardsticks end to end from here to a distant galaxy, the only measuring instrument we have for such distances is that provided by the paths of photons.

We can prove mathematically that only three possibilities exist for the geometry of space. Each possibility can be characterized by the curvature of space it entails (see Figure 6-3). Zero curvature corresponds to Euclidean geometry: Such a universe is called "flat." It extends forever, containing infinite volume and infinite mass. A negatively curved universe is also infinite in extent. We call these infinite models of the universe "open." By contrast, a positively curved universe has a finite volume and thus a finite mass (though it cannot have an edge if it is to satisfy the Cosmological Principle). A positively curved universe is called "closed."

We cannot visualize a curved volume since we lack a fourth spatial dimension from which to view it. Analogy must suffice, along with the equations that describe the geometry of a space of any number of dimen-

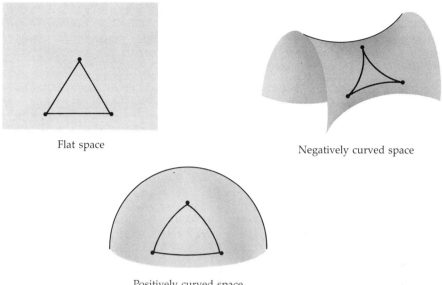

Flat space

Negatively curved space

Positively curved space

Figure 6-3. Two-dimensional analogues illustrate the three possible curvatures of three-dimensional space in our universe. Flat space is represented by an infinite plane, a portion of negatively curved space by the surface of a saddle, and positively curved space by the surface of a sphere. The triangles could represent the paths of photons sent between three galaxies at the corners of the triangles.

sions. What would the three types of curvature of the universe look like in a two-dimensional world? A zero-curvature universe would correspond to an infinite, flat sheet. A portion of a negatively curved universe would look something like a saddle, whereas a positively curved universe would correspond to the surface of a spherical balloon (see Figure 6-3). In a positively curved universe, a ray of light traveling in a straight line can eventually return to its source (though the universe will have evolved during the travel time). This corresponds to the fact that if we fly straight ahead from any airport on the earth, we shall eventually return to that airport after circumnavigating the globe. The same conclusion that applies to a two-dimensional closed surface (the surface of the earth or of a balloon) also holds true for a three-dimensional closed volume.

Let us illustrate one way to determine directly (in principle if not in practice) the geometry of our universe. Assume that we have neighbors in two distant galaxies and that our three civilizations have existed much longer than the millions of years required for signals to pass from one galaxy to another. Each civilization measures the angle between the positions of its two neighbors on the sky. In this way we determine the angle at each corner of the "cosmic triangle" formed by the photon paths connecting the three galaxies (see Figure 6-3). We can then determine the geometry of the universe simply by adding together these three angles. If the sum equals 180 degrees, the universe has zero curvature. If the sum is less than 180 degrees, the universe has negative curvature. If the sum is greater than 180 degrees, the universe has positive curvature. Once again, two-dimensional analogies prove helpful. If you draw triangles on the corresponding two-dimensional surfaces, you will obtain the same results.

The Expansion of the Universe

The large telescopes that probed the heavens during the early part of this century (1915–35) enabled astronomers to obtain the first estimates of the distances and the distribution of galaxies. The third major discovery made during this period proved even more remarkable, not least because it was almost totally unexpected. The prevailing belief, which followed naturally from the observation of apparently unmoving distant galaxies, was that the universe was unchanging—static and eternal. This belief was so strongly ingrained that even Albert Einstein felt obliged in 1917 to abandon the original and simplest version of his theory of gravitation one year after he produced it, because the theory did not allow for a static universe. Einstein felt compelled to add an additional term, called the cosmological constant, to his equations in order to balance the force of gravitation on the galaxies. Without this cosmological constant, the universe would be forced either to expand or contract, as was first demonstrated in detail by the Russian mathematician Alexander Friedmann in 1922.

The Spectra of Galaxies

Ironically, during the same time that Einstein was struggling to construct a theoretical model of a static universe, Vesto Melvin Slipher was obtaining the first evidence against such a model. Using the 60-centimeter refracting telescope at the Lowell Observatory in Arizona, Slipher obtained spectra of the light from various spiral galaxies. Since each spectrum provided a composite of the light from the billions of stars in a galaxy, each contained the characteristic absorption lines found in stellar spectra. But Slipher discovered a curious tendency. The spectral lines from most of the galaxies he examined had slightly longer wavelengths than those from nearby stars: They were "red shifted." Most astronomers ascribed these red shifts to the Doppler effect, caused by the motion of these galaxies away from us. Other suggested causes of the red shifts, such as the gravitational fields of the galaxies, proved incompatible with the observational evidence. (Radiation propagating out of a gravitational field becomes red shifted.)

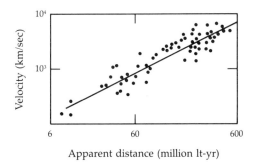

Figure 6-4. Observations of certain luminous spiral galaxies (each represented by a dot) show that their velocities (determined by their red shifts from the Doppler formula) are proportional to their distances (determined from their apparent brightness by the brightness formula, assuming they all have the same luminosity). The observations were made and analyzed by the astronomers Allan Sandage and Gustav Tammann at the Hale Observatories. (Redrawn from *Astrophysical Journal*, vol. 197, 1975, p. 272.)

Hubble's Law

As astronomers studied more galaxies and as they improved their estimates of galaxies' distances from us, another unexpected finding emerged. During the 1920s, Carl Wirtz, Knut Lundmark, and finally (and most convincingly) Edwin Hubble showed that the red shift of the light from a galaxy, and therefore the galaxy's velocity of recession, varies in proportion to the galaxy's distance from us. This relation between velocity (V) and distance (R), $V = H_o R$, is called "Hubble's law," and the constant of proportionality (H_o) is called the "Hubble constant" (see Figure 6-4).

This relation is, however, only approximate. All galaxies have a random component to their velocity, like the random velocity of the molecules in a gas, superimposed on their overall motion. Since this random velocity typically equals a few hundred kilometers per second, it exceeds the expansion velocity only for those galaxies closest to us. The simple velocity-distance relation discovered by Hubble also breaks down for extremely distant objects, where effects of relativity become important as the recession velocity reaches a significant fraction of the speed of light.

Hubble's profound conclusion still remains virtually inescapable today: The universe is expanding. At first glance this expansion might appear to violate the Copernican Principle, since we seem to occupy a preferred position, with all galaxies expanding away from *us*. But we do not occupy a preferred position, as we can understand by thinking of galaxies as molecules in a gas, confined to an imaginary box whose volume increases with time (see Figure 6-5). As the box expands, so does the gas. The

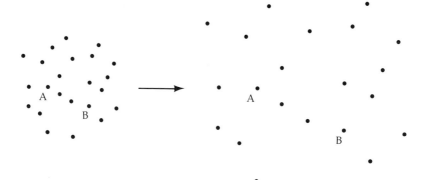

Figure 6-5. This drawing represents a sample of neighboring galaxies (not bound in a cluster), shown at two different times during the expansion of the universe. During this time interval, the average distance between the galaxies has doubled. Note that observers on any galaxy (such as A or B) see all other galaxies moving away from them. The change in the separation between any two galaxies during this time interval is proportional to their relative velocity (if this velocity is constant). This change in the separation of any two galaxies during the time interval must also be proportional to their initial separation, if the density of galaxies is to remain uniform in space. Hence the relative velocity between any two galaxies must be proportional to their separation (Hubble's law), since both are proportional to the change in separation within a given time interval. (We have ignored the small random velocities of the galaxies.)

Cosmological Principle requires that the density remain uniform in space, so the density must decrease with time as the volume grows. If you picture yourself on any molecule, you will see your neighboring molecules receding from you, each with a relative velocity proportional to its distance from you—no matter where you are. If you prefer a two-dimensional analogy, picture the galaxies as objects on an expanding rubber membrane (either an infinite sheet or a balloon). The same conclusions follow. Incidentally, each galaxy remains the same size, its constituents bound together by their mutual gravitational attraction.

The Age of the Expanding Universe

The proportionality between velocity and distance has another remarkable consequence: At some time in the past all the matter in the universe must have been extremely closely packed (at tremendously high density). If we mentally run the "movie" of the expanding universe backward in time, we can see that galaxies were once closer together. At some point in the distant past, the galaxies must have been touching. At still earlier times, the galaxies had no individual identities; their matter filled space as a uniform, high-density gas.

If we assume that the velocity of recession of each galaxy has not changed significantly during most of its lifetime, then the distance from us (R) that any given galaxy has traveled during that time is equal to its velocity (V) times its lifetime (t_o). This relationship ($R = Vt_o$) corresponds exactly to Hubble's velocity-distance relation ($V = H_oR$), if galaxies have a common age (t_o) that is equal to the inverse of the Hubble constant ($t_o = 1/H_o$). The quantity $1/H_o$ is thus a measure of the "expansion age of the universe." An observer in any other galaxy would obtain the same relation, and so the inverse of the observed value of the Hubble constant gives us a unique estimate of the time since the expansion "began."

Primeval Photons

This epoch of revolutionary cosmological discoveries ended almost as abruptly as it began. From the mid-1930s until the mid-1950s, astronomers and physicists learned much about the various types of stars and galaxies that comprise our universe, but they uncovered no new aspects of the structure of the universe as a whole. The reasons for this hiatus are not difficult to identify. The inauguration of the 5-meter Hale telescope in 1948 on Mount Palomar, near San Diego, only doubled the size of the largest telescope, although it allowed astronomers to see galaxies so distant that their recessional velocities reached about ⅓ of the speed of light. Most importantly, our view of the universe continued to be restricted to that narrow "visible-light" portion of the electromagnetic spectrum.

Greater and Greater Red Shifts

The next revolution in cosmology began in the 1950s, when astronomers began to observe the universe with radio telescopes. We learned in Chapter 4 what they found: huge jets of matter expelled from the cores of some galaxies. Radio telescopes soon became sensitive enough to allow astronomers to discover some radio sources at distances so great that the parent galaxy was too faint to be seen with ordinary (visible-light) telescopes. Still more interesting were radio sources that appeared to be starlike when viewed in visible light. These were the quasars, whose true nature began to be deduced in 1963.

The red shifts of the first quasars to be discovered were comparable to those of the most distant visible galaxies. Subsequently, however, astronomers have found hundreds of quasars with red shifts so large that the effects of general relativity modify the simple Doppler relationship between the red shift and the recession velocity.

In general, a source's red shift equals the difference between the received and emitted wavelengths of a spectral line, divided by the emitted wavelength. The largest red shift yet discovered (3.78) shifts the lines in this quasar's spectrum to wavelengths almost five times $(1+3.78)$ greater than those we would find from a nearby object. This quasar is at least 10 billion light-years from us (assuming that its red shift arises from the expansion of the universe), almost the maximum distance to which we can ever hope to see.

Astronomers cannot yet use quasars as accurate probes of the universe since quasars' distances cannot be accurately determined. If we knew the intrinsic luminosities of quasars, we could use the brightness relation to determine their distances, but we cannot reliably determine this luminosity (or most of their other intrinsic properties).

However, a quite different method of determining quasars' distances (and thus the value of the Hubble constant) may now be feasible because of the recent discoveries of gravitational lenses. Since the gravitational field of any massive object will bend a beam of electromagnetic radiation (see Figure 1-3), we can think of the gravitational field as acting like a lens. This effect can produce multiple images of a quasar if its light passes near any concentrated mass such as an intervening galaxy. These images correspond to certain allowed bending angles. Such a multiple quasar image was first discovered in 1979 (see frontispiece). If the radiation from the quasar varies with time, the difference in the arrival times of such signals in the various beams will allow astronomers to determine the distances to both the lens and the quasar and also to learn more about the distribution of matter responsible for the lens.

Astronomers have discovered another cosmologically significant property of the quasars. Many more quasars have large red shifts than we would expect from the number discovered with small red shifts. More quasars exist at large distances not because they are distributed nonuniformly through *space*, but because they are distributed nonuniformly in *time*. As we look farther out into space to the quasars with higher red shift, we look further back into the past, when quasars were more abundant.

The Cosmic Microwave Background Radiation

The most important discovery in cosmology made since the 1930s came in 1965, only 2 years after the discovery of large red shifts in quasars' spectra. Like the first observations of quasars by radio astronomers, this revolutionary discovery arose from the detection of long-wavelength electromagnetic radiation, but it occurred in an entirely different way, and emerged from an investigation whose intent was somewhat unrelated.

Physicists Arno Penzias and Robert Wilson at the Bell Laboratories in New Jersey were engaged in the development of an extremely sensitive microwave receiver (which detected radiation whose wavelength was 7.35 centimeters), designed to detect astronomical sources accurately. They found that they could not eliminate a troublesome source of background "noise," radiation that appeared to emanate from all directions on the sky. Curiously enough, a group at nearby Princeton University, headed by Robert Dicke, had just begun to look for such radiation. Dicke and another member of this group, James Peebles, quickly pointed out the possibility that the source of this radiation could be the universe itself, at an earlier epoch in its history. This indeed appears to be true. The significance of their discovery earned Penzias and Wilson the Nobel Prize in physics for 1978.

Twenty years earlier, the physicist George Gamow and his colleagues Ralph Alpher and Robert Herman had in fact predicted the existence of such microwave radiation. They based their prediction on their belief that most of the helium in the universe was synthesized by nuclear reactions during an early, hot phase of the expanding universe (as we shall see in Chapter 8). Alpher, Gamow, and Herman predicted that another remnant of this hot, dense, early phase of the universe is the radiation that today fills the universe as the microwave background. Oddly enough most scientists, including Penzias, Wilson, and the Princeton group, knew nothing of this remarkable prediction, which was contained in a few scientific papers published during the late 1940s.

The Spectrum of the Cosmic Background Radiation

As other groups began to study the microwave background radiation, two critical characteristics emerged that showed its origin must indeed be incredibly remote in space and time. When astronomers made measurements of the intensity of the radiation at various wavelengths, they found that its spectrum resembled that of blackbody radiation (see Figure 4-4). Blackbody radiation may be described as the radiation produced by an ideal radiator: matter that interacts strongly with radiation at all wavelengths. An example approximating an ideal radiator is the interior of a stove, whose walls emit and absorb photons of all wavelengths.

Recall (from Chapter 4) that the characteristic spectrum that emerges from an ideal radiator depends on only one property of the radiator, its temperature. Recall also that the radiation leaving the surface of a star has a spectrum similar to that of a blackbody (if we ignore the absorption and emission lines), characterized by a temperature on the order of 10,000 degrees Kelvin. But the radiation that Penzias and Wilson discovered has an intensity distribution characterized by a temperature of only 3 degrees Kelvin (see Figure 4-4 and Figure 6-6). Such a spectrum peaks at a wavelength of about 2 millimeters, 3,000 times greater than the average wavelength of visible radiation. Where is the matter that produced this radiation? The answer appears, at least in part, in the second critical property of the microwave radiation.

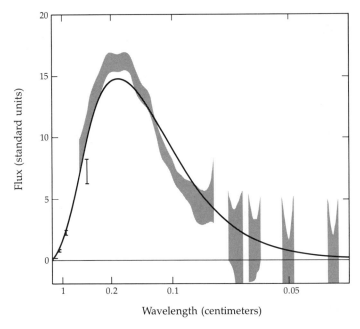

Figure 6-6. This diagram shows the observed spectrum of the microwave background radiation, including both the early ground-based measurements (indicated by vertical error bars) and the recent measurements (indicated by shaded areas) carried out from a balloon by David Woody and Paul Richards of the University of California at Berkeley. The theoretical blackbody spectrum (smooth line) corresponds to a temperature of 2.96 K. (Redrawn from *Astrophysical Journal*, vol. 248, 1981, p. 28.)

The Isotropy of the Cosmic Background Radiation

Penzias and Wilson found that the intensity of the radiation they discovered appeared to be the same in every direction they pointed their antenna. Subsequent measurements by other observers confirmed this isotropy of the microwave background with increasing accuracy. Such isotropy could arise if we were surrounded by enough matter to absorb and emit the radiation, but we know that there is not enough interstellar matter in our galaxy to do this. A uniform distribution of many sources of this radiation over the sky could also produce the observed isotropy. However, our galaxy certainly looks quite different as we view it in various directions, and the contribution of nearby galaxies would certainly not be isotropic. In addition, we know of no objects within galaxies capable of producing blackbody radiation characterized by such a low temperature.

The conclusion that emerged with increasing force was this: The microwave radiation that fills the sky was produced in the remote past, when the gas that then filled the universe had sufficient density to behave like an ideal radiator—frequently scattering, absorbing, and emitting photons. We can also think of this radiation as a uniform gas of photons that has expanded with the universe and has cooled in the process.

This picture of the source of the microwave background implies that the radiation should not be strictly isotropic. Galaxies have random velocities, equal to about 0.1 percent of the speed of light. Therefore galaxies should be moving through the sea of microwave photons in various directions with such speeds. Our own Milky Way galaxy, likewise adrift in the cosmic sea of photons, thus should have a similar velocity of several hundred kilometers per second with respect to this sea.

The Doppler effect of velocity on wavelength must apply to the photons we observe in the microwave background. In the direction of our motion, we detect radiation toward which we are moving. This radiation therefore appears shifted toward shorter wavelengths and higher energies (see Figure 6-7). Hence its intensity is increased somewhat. Conversely, radiation that we receive from directions opposite that of our motion through the photon sea must appear shifted toward longer wavelengths and lower energies. Its intensity is therefore decreased somewhat. Thus an apparent anisotropy (departure from isotropy) results from our motion. The Doppler formula relating the shift in wavelength to velocity tells us that this anisotropy—the relative variation in the intensity of the background radiation over the sky—should be approximately equal to the speed of our galaxy (its random velocity) divided by the speed of light.

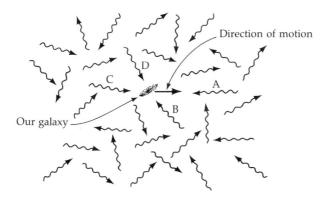

Figure 6-7. This sketch illustrates the motion of our galaxy through a portion of the gas of photons that forms the cosmic microwave background radiation. Radiation that we receive from directions toward which we are moving (photons A and B) is shifted to shorter wavelengths because of the Doppler effect, which increases its intensity. Correspondingly, radiation that we receive from directions opposite to our motion (photons C and D) is Doppler shifted to longer wavelengths, which decreases its intensity.

Physicists George Smoot, Marc Gorenstein, and Richard Muller at the University of California at Berkeley and Brian Corey and David Wilkinson at Princeton University confirmed the existence of such anisotropy in 1977. The observed degree of anisotropy implies a random velocity for our galaxy equal to 0.2 percent of the speed of light. This result conclusively demonstrates that the radiation arises in regions beyond our galaxy. (If the radiation came from our galaxy it could not exhibit the particular type of anisotropy observed.) When we allow for the motion of our galaxy, the microwave background radiation appears to be intrinsically isotropic to an accuracy of at least 0.01 percent.

We have now discussed the basic facts about the universe that have been revealed as we have extended the horizons of our observations. To give these facts meaning and to produce a comprehensive picture of the universe, we must develop a theoretical model. This task now awaits us in the next chapter.

7

The Big Bang Model of the Universe

IN THE PREVIOUS FOUR CHAPTERS, we have surveyed the present universe. On a fundamental level, we have found that it consists of three types of elementary particles: quarks, leptons, and bosons. The tremendous variety of structures, ranging in size from nucleons to clusters of galaxies, arises from the ways these particles bind together through three fundamental interactions—the strong, electromagnetic, and gravitational forces. The bulk of the matter in the universe has one of three forms: "ordinary" visible matter (composed of protons, neutrons, and electrons), such as stars and gas; "invisible" matter (whose nature is unknown), revealed solely by the gravitational force it produces; and radiation (composed of photons and possibly other massless particles), most of which appears in the microwave background. We have also discussed the compelling evidence that the totality of this matter, our universe, is expanding.

Evidence, Assumption, and Extrapolation

We now confront a major task of the theorist—to convert our knowledge of the present universe into a picture of its past and future evolution. We accomplish this extrapolation in the same way that we solve any dynamical problem in physics: with the use of the equations of motion that govern the system under study. Because the gravitational force controls the motion of all large aggregates of matter, including the universe, a comprehensive theory of gravitation must provide the equations that govern our model. If we can determine certain key properties of the universe in its present state, the equations governing these properties then allow us to determine the past and future of the universe. To do this we must also know the forms of matter that existed in the past and that will exist in the future. This analysis resembles the calculation of the past and future path of a ball by observing the position and velocity of the ball at one instant of time.

In constructing such a model of the evolution of the universe, we follow the approach that has proven best in other areas of physics. Our key requirement is simplicity: We construct the simplest model (mathematically speaking) that agrees with all observations. As with any theoretical model, as soon as any prediction based upon this model fails to agree with any observation, we must abandon the model in favor of a new one.

This simplest model of the universe is the so-called big bang model, which has been adopted by most cosmologists today. It originated in the 1920s with the work of Alexander Friedmann and the Belgian cleric Georges Lemaître, who independently discovered the simplest family of solutions to Einstein's equations of general relativity that can describe our universe. In the big bang model, the universe has expanded from a state in which all the matter was compressed to extremely high density at a particular time in the past. It is important to note that the "explosion" from this state did not occur in a single region of space, but rather encompassed *all* points of space at a certain instant of time.

Four basic assumptions define the standard big bang model of the universe:

1. The theory of gravitation that governs the evolution of the universe is general relativity. General relativity, as originally formulated by Einstein in 1916, is the mathematically simplest theory of gravitation that agrees with all experiments and observations. It incorporates the theory of special relativity, formulated by Einstein in 1905, which deals with phenomena not involving gravitation.

Einstein based both theories upon the view that space and time do not have independent significance, but rather are related by transformations within the four dimensions of *space-time*. An event is a point in space-time, with a position described by three spatial coordinates (where it happened) and one time coordinate (when it happened). All observers find the laws of physics to be the same, irrespective of the particular spatial and time coordinates that the observers choose. In particular, all observers find the speed of all passing massless particles to be the same: the speed of light. Particles that move at or near the speed of light are called *relativistic*, whereas those that have speeds much less than that of light are called *nonrelativistic*. Newton's laws of motion and gravitation become good approximations to those of Einstein only when matter moves nonrelativistically and experiences weak gravitational forces (such as the gravitational field of the earth).

Evidence supporting Einstein's theory of general relativity has accumulated over the years, especially during the 1960s and 1970s. Astronomers have measured the paths of the moon and the planets, as well as the paths of spacecraft, with an accuracy sufficient to rule out most competing theories. Because photons, like all forms of matter, are affected by gravitation, any beam of electromagnetic radiation bends as it passes the sun (or any other massive object), as we saw in Figure 1-3. In addition, the sun's gravitational field produces a slight increase in the time required for any photon to pass by the sun, compared with the time required if the sun were not there. Measurements of these two effects, bending and time delay, have provided additional confirmation of Einstein's theory. In the last few years, astronomers have found that the two members of a particular binary star system, one of which is a pulsar, are slowly spiraling toward each other at a rate that agrees with the rate expected from the loss of energy through gravitational radiation predicted by Einstein's theory (see Chapter 3). This agreement between observation and theory provides the most powerful test of all, for it depends in a sensitive way on the precise structure of general relativity theory.

2. The standard big bang model also assumes that the Cosmological Principle is valid, at least over distance scales comparable to the size of the visible universe at any epoch. Recall that this principle posits that the properties of the universe appear the same when viewed by any observer who moves with the universal matter, whatever the direction in which he looks (isotropy) or wherever he is located (homogeneity).

We saw evidence in Chapter 6 supporting this assumption for regions of the present universe large enough to contain many galaxies. However, as we shall see, in the distant past the visible universe surrounding any observer probably contained far less matter. This assumption therefore requires that in the past, matter was likewise distributed uniformly on these smaller scales of the then-visible universe, becoming nonuniform only at later times. This picture of a universe that becomes more "lumpy" with time is in fact expected within the standard big bang model, as we shall see.

3. The present universe contains virtually no antimatter in the form of antineutrons, antiprotons, or positrons; it may, however, contain other types of antiparticles, such as antineutrinos. (We reviewed the evidence supporting this assumption in Chapter 4.) According to the big bang model, the composition of the universe was quite different in the remote past. The early universe was so hot that all types of antiparticles existed in abundances almost equal to those of their corresponding particles.

4. At all epochs, the great bulk of the mass of the universe consists of the known types of elementary particles. This assumption specifically precludes the possibility that "empty" space—a region containing no particles—contains mass, and therefore gravitates. Hence it rules out the presence of the "cosmological constant" in Einstein's equations (see Chapter 6), since this term allows empty space to produce gravitational effects. It also rules out fields that can be described not only in terms of particles

but also in terms of a contribution by the field to the mass of empty space. We shall consider an example of this latter possibility in Chapter 8.

These four assumptions, which together form the foundation upon which the standard big bang model rests, have immense power. If valid, they allow us to extend our knowledge of the universe to epochs in which the physical conditions (such as temperature, density, and types of particles present) were far more extreme than any we can produce in terrestrial laboratories, or that nature can produce anywhere in the presently observable universe. This extension in time of our picture of the universe represents one of the greatest extrapolations in all of science. The power and importance of these assumptions therefore impose a special obligation on us to subject them to every possible observational test.

The Dynamics of the Universe

Einstein's theory of general relativity plays a dual role in describing any system (including the universe) in which gravitation is the dominant force. The theory not only describes how matter "curves" space-time, the geometrical description of a gravitational field, but also describes how this curvature in turn acts on matter, determining how the matter will move.

A two-dimensional analogy illustrating this dual nature of general relativity consists of a large steel ball (representing, for example, the sun) placed on a rubber membrane, making the membrane curve. (The curvature represents the gravitational field of the sun.) The weight of the ball curves the membrane; the shape of the membrane in turn determines how other balls will move (see Figure 7-1). A smaller steel ball (representing a planet) orbits the larger ball when placed in motion on the curved membrane. The curvature of the membrane decreases with distance from the large ball, just as the gravitational field of the sun decreases with distance from it.

We have discussed the possible three-dimensional large-scale curvatures of space allowed by the Cosmological Principle (see Chapter 6). However, since relativity teaches us that the laws of nature are properly expressed within the four dimensions of space-time, the general description of gravitation involves the curvature of the geometry of space-time. Nevertheless, the spatial curvature still plays an important role in the full four-dimensional description of the universe that we obtain from Einstein's equations.

A Sample of the Universe

Let us now apply Einstein's theory of general relativity to the universe as a whole. To do this, it is sufficient to focus on an imaginary sphere with our galaxy (or any other) at its center. This sphere has a radius just large enough to contain a representative sample of the universe (many galaxies), as shown in Figure 7-2. The radius of the sphere must change in such a

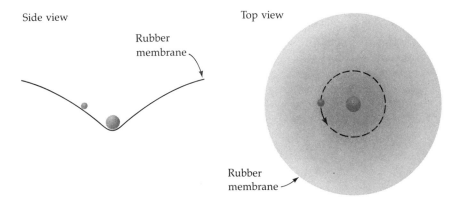

Side view

Top view

Rubber
membrane

Rubber
membrane

Figure 7-1. The effects of gravitation can be thought of in terms of the curvature of space-time, analogous to the curvature of a two-dimensional membrane produced by the weight of a large steel ball. The curvature of the membrane increases as one approaches the large ball, just as the gravitationally induced curvature of space-time increases as one approaches a massive object. The small ball will be guided in an orbit around the large ball by the membrane. Similarly, the planets are guided in their orbits about the sun by the curvature of space-time produced by the sun.

way that the total amount of ordinary matter (mostly in the form of neutrons and protons) within the sphere remains constant in time as the universe expands or contracts.

When we examine much earlier phases of the universe, during which the sphere contained a hot gas of various types of particles, we must be more precise: The total *baryon number* of all the matter within the sphere must remain constant. The baryon number equals ⅓ for each quark, −⅓ for each antiquark, and therefore equals 1 for neutrons and protons, −1 for antineutrons and antiprotons, 0 for mesons (quark plus antiquark), and 0 for leptons and bosons. To obtain the total baryon number, we simply add (noting the sign) the baryon numbers of all particles present. The total baryon number is conserved (remains constant) to a high degree of accuracy in all known reactions among particles, although this conservation law may be violated under some circumstances (as we shall discuss in Chapter 8). For now, we can use the total baryon number in our imaginary sphere as a means of following a given, arbitrary amount of matter during the evolution of the universe.

Because our imaginary sphere is imbedded within the actual distribution of matter, its boundary must move at the same velocity (V) as does the matter in its vicinity. (The larger the radius [R] of the sphere, the larger the velocity, as Hubble's law indicates.) Each galaxy (or particle) on the sphere's boundary feels a gravitational force (F) that can be thought of as

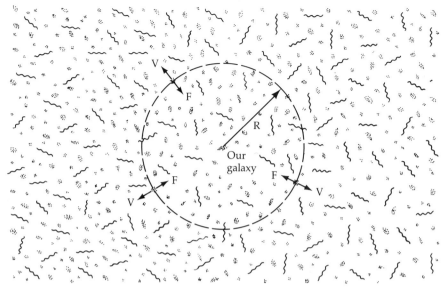

Distance (R) = present distance (R_o) × scale factor (S)

Velocity (V) = expansion rate (H) × distance (R)

Mass-energy (M) = mass-energy density (ρ) × volume ($\frac{4}{3}\pi R^3$)

Figure 7-2. A representative sample of the universe resides within the (imaginary) sphere drawn here. The sphere is large enough to include many galaxies—at least several thousand. Within this sphere of representative matter, the total baryon number (neutrons plus protons) remains constant. Radiation is denoted by the wavy lines. The arrows represent the directions of the velocity (V) of, and the gravitational force (F) on, galaxies on the surface of the sphere. Equations for distance (R), velocity (V), and mass-energy (M) are also given.

being produced by all the matter within the sphere on which the particular galaxy is located. The matter outside the sphere, pulling in various directions, produces no net gravitational force on a galaxy (or particle) on the boundary of the sphere.

A major difference between Newton's theory and Einstein's is that in Einstein's theory *all* forms of mass and energy, including radiation, contribute in the same way to the total gravitational field, whereas in Newton's theory only the ordinary mass of particles contributes. According to Einstein, the *mass-energy* (M) of all particles provides the source of gravity. Mass-energy is the sum of the ordinary (rest) masses (m) of particles and their kinetic masses. (The kinetic mass of a particle equals its energy of motion [kinetic energy] divided by the square of the speed of light [c].) The amount of energy equivalent to mass-energy is the sum of the rest-mass energy (mc^2) and the kinetic energy of the particles.

Another difference between the two theories is that in Einstein's theory the pressure of matter, as well as its density of mass-energy, contributes to the gravitational force. In Newton's theory, only density contributes.

In our model we assume that matter has a uniform (homogeneous) distribution. Hence no forces caused by pressure gradients can arise since the pressure is the same everywhere at any given time. A star remains in equilibrium, neither collapsing nor expanding, because the inward gravitational force on its gas atoms balances the outward force produced by the decrease in pressure from the star's center to its surface (see Chapter 5). The universe, by contrast, contains no pressure forces to balance the force of gravitation over the large distances involved. Hence no equilibrium state can exist. The universe must be dynamic, always either expanding or contracting. The only exception could be a momentary transition from expansion to contraction, or vice versa.

In this model, the distance between any two representative particles in the universe is proportional to what we call the "universal scale factor" (S). The scale factor describes how the separation between any two such neighboring particles in the universe varies with time: S equals the distance between two particles at a certain time (R) divided by their separation at the present time (R_o), as indicated in Figure 7-2. Hence S equals one at the present epoch. Since the universe is expanding, S was less than one in the past and will be greater than one in the future. It is important to note that this scale factor depends only on time, not on position in the universe. The separations of all particles in the universe change in the same way with time.

The Cosmological Principle requires that each part of the universe evolve in the same way. When we apply Einstein's theory of general relativity to a typical region, such as that shown in Figure 7-2, two equations emerge. The first, which we shall call the evolution equation, governs the dynamics of the universe. The second, which we shall call the mass-energy equation, governs how the total mass-energy within our imaginary sphere changes with time.

Einstein's Evolution Equation

Einstein's equation of universal evolution describes how the rate at which the universe expands or contracts (H) is determined by the total density of mass-energy (ρ) and the present curvature of space (κ). The equation is: $H^2 = \frac{8}{3}\pi G\rho - \kappa c^2/S^2$, where G is the gravitational constant. Figure 7-3 shows how the two terms on the right-hand side of this equation vary as the universal scale factor evolves.

The first term on the right-hand side of the equation, the density term ($\frac{8}{3}\pi G\rho$), is proportional to the mass-energy density ρ, which decreases as the separation between particles (and therefore S) increases. The second

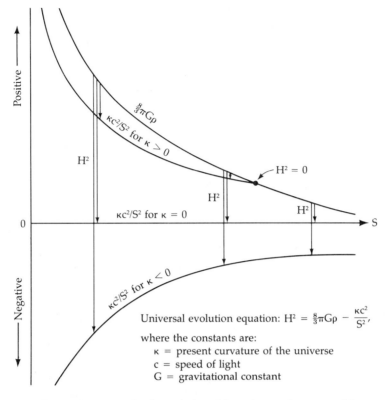

Figure 7-3. In Einstein's equation for the evolution of the universe, the square of the expansion (or contraction) rate (H) equals the difference of two terms, one proportional to the average density of matter (ρ) and the other proportional to the present curvature of the universe (κ) divided by the square of the scale factor (S). The graph shows how each of the two terms varies as the scale factor changes; the curvature term is shown for different values of the present curvature (positive, zero, and negative). The length of the arrows indicates the difference between the two terms, which gives the value of H^2 for each choice of κ. At the point $H^2 = 0$, the expansion of a closed universe ($\kappa > 0$) reverses to contraction.

term, the curvature term ($\kappa c^2/S^2$), is proportional to the curvature of space. Its behavior depends upon whether the curvature is positive, negative, or zero, as shown by the three curves plotted in Figure 7-3. The curvature term also changes with time, since it is inversely proportional to the square of the scale factor. The universal evolution equation states that the density term minus the curvature term equals the square of the expansion (or contraction) rate. All terms in this equation depend only on time, not position. The Hubble constant, H_o, is the present value of H.

We now proceed to the second equation provided by general relativity theory, the mass-energy equation, which predicts how the total mass-energy within our imaginary sphere changes with time as the universe evolves. When most of the matter has random velocities far less than that of light (apparently the situation in the present universe), most of the mass-energy of the matter (e.g., galaxies, stars, and atoms) resides in ordinary rest mass. In this case the equation states that the total mass-energy (M) within the sphere, which equals the product of the density of mass-energy (ρ) and the volume of the sphere (proportional to R^3, and therefore to S^3), remains constant as the universe evolves (see Figure 7-2). In other words, the product ρS^3 remains constant if the universe is dominated by nonrelativistic matter.

A different consequence of the mass-energy equation follows if most of the matter in the universe is relativistic. This was very probably the case in the early universe, which was filled with massive particles with speeds close to that of light, as well as with massless particles, whose speed is always equal to that of light. Like photons, all such relativistic particles can be described in terms of waves. The wavelength characterizing a relativistic particle is inversely proportional to its energy, which takes the form of kinetic energy. A relativistic particle's rest-mass energy is either zero (for massless particles) or negligible. The wavelength of such a particle expands or contracts with the universe and is therefore proportional to the scale factor. (This description provides one way to understand the enormous red shifts of the microwave background radiation produced by the expansion of the universe.) The energy (E) of a relativistic particle is thus inversely proportional to the scale factor. As the scale factor increases, the particle's energy decreases, and vice versa.

However, as in the case of ordinary, nonrelativistic particles, the total number of such relativistic particles within our imaginary sphere does not change with time. The number of particles per unit volume (their number density [n]) times the volume of the sphere (which is proportional to S^3) thus remains constant. The particles' mass-energy density (ρ) equals their average individual mass-energy (E/c^2) times their number density (n): $\rho = En/c^2$. Therefore, since E is inversely proportional to S and n is inversely proportional to S^3, ρ must be inversely proportional to S^4. In other words, the product ρS^4 remains constant if the universe is dominated by relativistic matter.

The difference between the behavior of nonrelativistic matter (ρ inversely proportional to S^3) and relativistic matter (ρ inversely proportional to S^4) arises solely from the fact that the total energy (mostly in the form of rest-mass energy) of a nonrelativistic particle cannot change, whereas

the total energy (mostly in the form of kinetic energy) of a relativistic particle is reduced by the expansion of the universe. The fact that the mass-energy density of relativistic particles falls more rapidly than that of non-relativistic particles as the universe expands (S increases) merely reflects the effect of the red shift on the energy of relativistic particles.

What the Equations Reveal

Armed with the different behavior of the mass-energy density ρ as the scale factor changes during epochs dominated by either nonrelativistic or relativistic matter, we can now return to Einstein's evolution equation: $H^2 = \frac{8}{3}\pi G\rho - \kappa c^2/S^2$ (see Figure 7-3). For either of the two cases, the density term $\frac{8}{3}\pi G\rho$ decreases rapidly as the scale factor S increases. The evolution of the universe then emerges if we follow how the difference between the density term and the curvature term $\kappa c^2/S^2$ changes as the scale factor changes for a particular choice of curvature κ. This difference, equal to the square of the expansion (or contraction) rate, H^2, is shown by the arrows in Figure 7-3. For each choice of curvature, the varying length of the arrow shows how H^2 varies as S changes with time. As S increases while the universe expands, the length of the difference arrow (H^2) decreases for any possible curvature. If the curvature is positive (a closed, finite universe), the difference arrow eventually vanishes (at the point where the density and curvature terms are equal). Hence the expansion must cease ($H = 0$) at that value of S. The universe will then begin contracting, with the difference arrow moving back to the left. By contrast, if the curvature is zero or negative (open, infinite universes), the expansion must continue indefinitely, with the expansion velocity (V) of any particle eventually approaching zero (if the curvature is zero) or a constant (if the curvature is negative).

Figure 7-4 illustrates these possibilities in another way by showing the motion of a single galaxy (or particle), at a particular distance from us today, for the three types of curvature. Distance from us is plotted versus time. The slope at any point on each curve (the rate of change of distance with time) represents the galaxy's velocity at that time. Each curve has the same slope today (at point A), since that slope, divided by the galaxy's distance, equals the present value of the expansion rate, H_o, the Hubble constant.

By evaluating the terms in the evolution equation at the present epoch and dividing each term by $\frac{8}{3}\pi G$, we obtain the equation $\rho_o - \rho_c = 3\kappa c^2/8\pi G$. The present density of the universe is ρ_o, while the critical density ρ_c is defined as equal to $3H_o^2/8\pi G$. If the present density is greater than the critical density, we see that the curvature must be positive, and hence the universe must eventually contract. This seems reasonable, since a greater density of matter implies a greater gravitational force. By contrast, if the

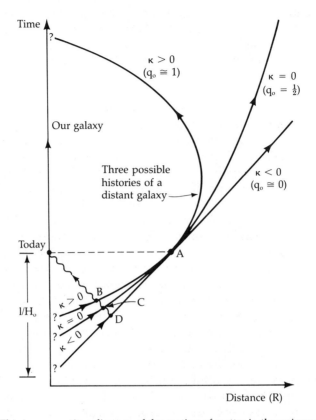

Figure 7-4. This is a space-time diagram of the motion of matter in the universe. The motion of a galaxy relative to a typical observer (for example, us) depends on whether the universe has positive, negative, or zero curvature. The galaxy's distance from us is plotted as a function of time for the three possible types of curvature. The history of our own galaxy is represented by the vertical line at zero distance. The galaxy's present position is shown at point A. Notice the different futures predicted for different curvatures: Positive curvature means eventual contraction, whereas negative and zero curvature mean eternal expansion. The corresponding values of the deceleration parameter (q_0) are also given (in the case where the present pressure of the universe is negligible). The wavy line shows the path through space-time of those photons emitted by the galaxy in the past (at either point B, C, or D) that reach us at the present.

present density is equal to or less than the critical density, the curvature must be zero or negative, and therefore the universe must expand forever. One of the greatest observational challenges cosmologists now face consists of measuring the present density and the critical density (via the Hubble constant) with sufficient accuracy to determine the present curvature—and thus the fate—of the universe.

Measuring the Density

As we saw in Chapter 4, we can determine the part of the total present density that resides in gravitationally bound systems with some visible matter. (We do this by measuring the size and internal velocities of such systems.) Any contribution to the present density that is distributed uniformly throughout the universe (such as the microwave background photons, since they cannot clump by gravitation) can also be determined *if* we can detect it directly. However, the only reliable way to determine the *total* present density of the universe is to measure the gravitational effect of all the mass-energy.

We can make this measurement because the mutual gravitational attraction of the matter in the universe tends to slow its expansion. Therefore the density of the universe governs its deceleration: the rate at which its expansion velocity decreases. Cosmologists express the present deceleration in terms of a quantity q_o, the deceleration parameter. The gravitational force, and therefore the deceleration parameter, is actually proportional to the sum $\rho_o c^2 + 3p_o$, where p_o denotes the present pressure in the universe. However, since the present pressure is probably negligible, if we measure the deceleration parameter, we determine directly the present total density of mass-energy, ρ_o.

Measuring the Deceleration

One way to determine the deceleration of the universe can be seen from Figure 7-4. There we see the history of photons emitted by a particular galaxy in the distant past and received by us today. Notice that the distance of the galaxy from us at the time it emitted the light we see today (the event indicated by point B, C, or D) depends on the deceleration of the universe (the rate of change of the slope of the curves). The red shift of any spectral lines will also depend upon the deceleration, since the wavelength of all photons will be "stretched" by the expansion of the universe in proportion to how much the universe has expanded during the photons' travel time to us. This stretching factor equals 1/S, the ratio of the distance of the galaxy from us today (R_o) to its distance from us when it emitted the photons we receive today ($R_o S$). (Recall the formula in Figure 7-2.)

As we learned in Chapter 4, an observed property of an object can provide us with a measure of the object's distance from us (related to the distance shown in Figure 7-4) if we know the corresponding intrinsic property of the object. Each such measure of the distance of an object of known red shift is related to the value of the Hubble constant H_o, as well as to the value of the deceleration parameter q_o. The dependence on H_o can be eliminated, and the value of q_o obtained, by comparing observations of

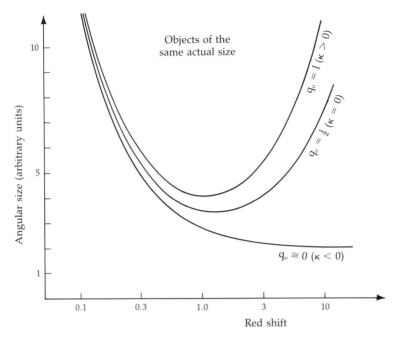

Figure 7-5. This diagram depicts how the observed (angular) size of an object with fixed actual (linear) size depends upon its red shift. The relation is plotted for three choices of curvature (κ), corresponding to three values of the deceleration parameter (q_o) as in Figure 7-4. The red shift equals the increase in the wavelength of radiation as it travels from the source to us divided by the wavelength emitted by the source. An object's red shift is proportional to its distance only when the red shift is much smaller than one.

distant objects with those of nearby objects whose relevant intrinsic property, such as luminosity or size, can be shown to be the same.

For example, Figure 7-5 shows how the observed angular (apparent) size of objects with the same actual linear (intrinsic) size depends on the red shift of light received from the objects. Note that for small red shifts (which correspond to nearby objects) we obtain the expected result: The apparent size of an object decreases as its distance increases. But for large red shifts we obtain quite a different behavior, which in addition depends on the value of the deceleration parameter. In fact, the observed angular size of objects of the same actual size eventually increases with increasing red shift. This reflects the fact that red shift is no longer proportional to distance.

We also note from Figure 7-4 how the magnitude of the deceleration is related to the curvature of space in the universe. Let us continue to assume that the present pressure in the universe is negligible. Then if the deceleration parameter q_o is greater than $\frac{1}{2}$, the present density ρ_o must be greater than the critical density ρ_c, the curvature of the universe must be positive, and the universe must eventually contract. If the deceleration parameter q_o is less than ½, the present density ρ_o must be less than the critical density ρ_c, the curvature must be negative, and the universe must expand forever. The borderline case $q_o = ½$ corresponds to ρ_o equal to ρ_c and to zero curvature, with the universe expanding at an ever-decreasing rate.

In Chapter 4, we also discussed some of the difficulties in distance determination that arise from the necessity of determining the intrinsic properties of an object by indirect means. Here, in dealing with extremely remote objects, we encounter an additional problem, which arises from the need to compare the properties of objects at different cosmic epochs. To carry out this comparison, we must know how the intrinsic property used in a particular method of distance determination evolves with cosmic time. The difficulty of computing this correction for evolution produces the major uncertainty in our present determination of the deceleration parameter. The magnitude of this uncertainty still allows for positive, zero, or negative curvature. But the uncertainty is small enough to tell us that the present density of all forms of matter in the universe cannot be more than a few times greater than the critical density.

A Newtonian Analogy

The three possibilities for the evolution of the universe depicted in Figure 7-4 have their counterparts in ordinary Newtonian mechanics—for example, in the motion of a projectile fired from the earth's surface. If the velocity acquired by the projectile exceeds a critical value (the escape velocity), then the projectile will never return to earth (see Chapter 5). Its velocity will simply decrease to some final, constant value as the result of the deceleration produced by the earth's gravitational field. This corresponds to the case of a universe with negative curvature, which will expand forever. If the initial velocity precisely equals the escape velocity, the projectile's speed will decrease toward zero at great distances from the earth. This corresponds to a universe with zero curvature, in which the relative expansion velocity of all matter will decrease toward zero as time extends to infinity. If the projectile's initial velocity is less than the escape velocity, the projectile will eventually fall back to earth. This corresponds to a universe with positive curvature, which will eventually contract. (Note that the galaxy in Figure 7-4 will eventually begin to approach us if the curvature is positive.) In all cases, the properties of the projectile's motion are determined by its initial velocity and by the strength of the gravitational force on it. Correspondingly, the properties of the universe's motion are determined by its expansion rate and density today (or at any other particular time).

When Did the Expansion Begin?

The final universal property illustrated in Figure 7-4 is the "age" of the universe. Note that the galaxy whose motion is depicted must have been extremely close to us (its distance R almost zero) at a certain time in the past. At such early times the universe actually consisted of a nearly uniform gas, whose irregularities grew to form galaxies. Hence Figure 7-4 represents the history of any bit of matter contained within that galaxy today, or, more broadly, it tells us how the distance between any two particles in the universe changes with time. We see that any two such particles must have been much closer to each other at some time in the past than they are today. Roughly speaking, this was the time of the big bang. Moreover, the Cosmological Principle ensures that this time of proximity must be the same for any pair of particles. Otherwise the density of particles (which approaches infinity as the distances between particles approach zero) would not remain uniform in space as we examine the past. However, we see that this time of proximity (as measured back from today) depends upon both the value of the Hubble constant and the value of the deceleration parameter (or, equivalently, the curvature).

The particular curve $\kappa < 0$ representing negative curvature in Figure 7-4 depicts the limiting case in which the curvature term now far exceeds the density term on the right-hand side of the evolution equation. This dominance of the curvature term produces an almost constant expansion velocity (hence a straight line on the plot of distance versus time, corresponding to small deceleration, $q_o \cong 0$); the expansion began at a time in the past that approximately equals the inverse of the Hubble constant. From measurements of H_o, we can determine that this time ($1/H_o$) has a value between 10 and 20 billion years. As we consider universes with curvatures increasing from negative values through zero to positive values, the deceleration becomes greater, resulting in model universes that began their expansion at somewhat more recent times than $1/H_o$, and in which the expansion velocity changed more significantly as the universe evolved (as shown in Figure 7-4).

Does this time since the expansion began equal the age of the universe? We do not know, because we do not know the properties of matter and gravitation when the density and temperature had incredibly high values in the distant past. We therefore place question marks in Figure 7-4 where the distance between particles approaches zero. Was matter "created" at a finite time in the past? Did the universe "bounce" from a previous phase of contraction, with the particle separations never becoming zero? We shall discuss such remote epochs in Chapter 8; for now we shall confine our attention to times when the density of matter was low enough to allow us to describe its properties with confidence.

The Recent History of Matter

With this overall survey of the big bang model of the universe in mind, let us now look in more detail at some of the consequences of our theoretical scenario. In the remainder of this chapter we shall consider that portion of the history of the universe accessible to our direct view; that is, epochs from which photons can propagate to us. More remote epochs, forever shrouded from our view by the high density of particles present at those times, will be discussed in Chapter 8.

Figure 7-6 provides a view of what the universe may have been like during five "recent" epochs, ranging from 10,000 to 10 billion years after the expansion began. We focus on a portion of the universe containing an amount of ordinary matter equal to that contained in a few galaxies. Figure 7-6 shows the distribution of matter, as well as the time (t) since the expansion began, the universal scale factor (S), and the radiation temperature (T) at five different epochs, with the universe expanding by a factor of ten between each epoch. Hence each picture of the universe should be imagined as ten times larger than the one preceding it. (Recall that the time scale is uncertain by a factor of at least two, whereas the radiation temperature is known with relative accuracy.)

Cosmologists believe that during such recent epochs, the basic constituents of the universe have been nonrelativistic matter (electrons, nuclei, and atoms, plus whatever particles comprise the invisible matter), along with radiation (photons), and undetectable neutrinos and antineutrinos (which may comprise the invisible matter). Most of the radiation in the universe does not emanate from specific sources, as starlight does. Instead it comes to us with equal intensity from all directions, in the form of the microwave background radiation. The source of this radiation was the entire universe at an early epoch.

This sea of microwave photons contributes very little to the present mass-energy density of the universe—no more than one part in 10,000. Recall that relativistic particles such as photons contribute their kinetic masses to the total mass-energy density, whereas nonrelativistic particles contribute primarily their rest masses. As we look back in time, the contribution of the photon sea to the mass-energy density grows in importance relative to the contribution of ordinary, nonrelativistic matter, since each photon had higher energy then. Because each photon's energy is inversely proportional to the scale factor, the photons' mass-energy density exceeded that of ordinary matter at epochs earlier than the first one depicted in Figure 7-6. Nowhere in the present universe do photons dominate the mass-energy density; most of the mass-energy of even the hottest stars arises from ordinary, nonrelativistic matter, not from the radiation within the stars.

Expansion time (t) (years)	Scale factor (S)	Radiation temperature (T) (degrees Kelvin)		
10,000	1/10,000	30,000	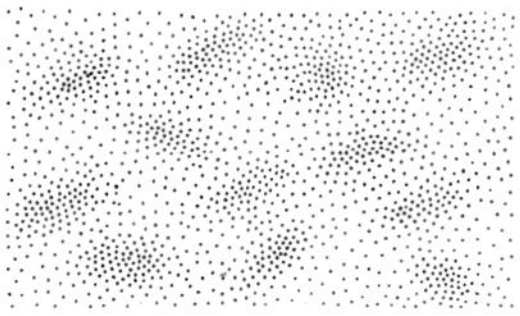	Primeval plasma
300,000	1/1000	3,000	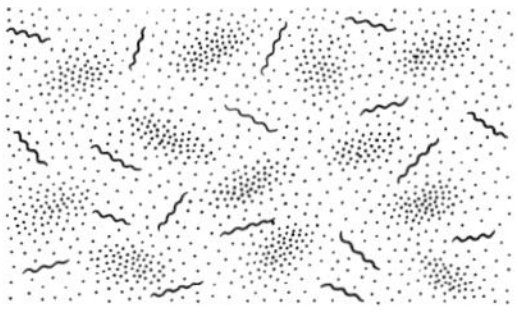	Atoms form
10,000,000	1/100	300	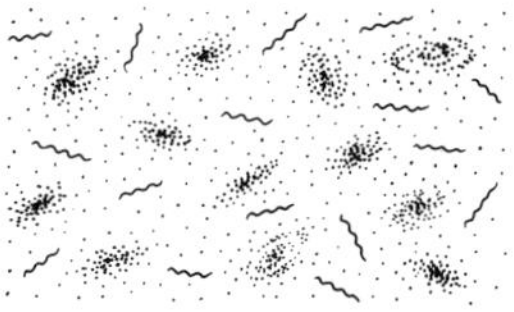	Galaxies form
300,000,000	1/10	30	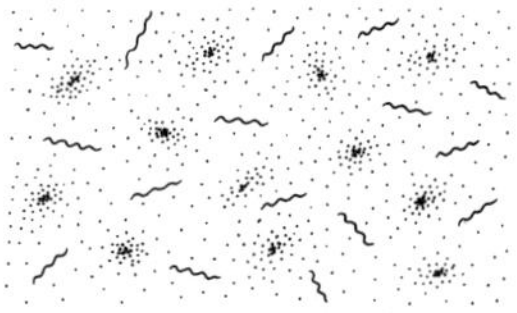	Galaxies collapse
10,000,000,000	1	3	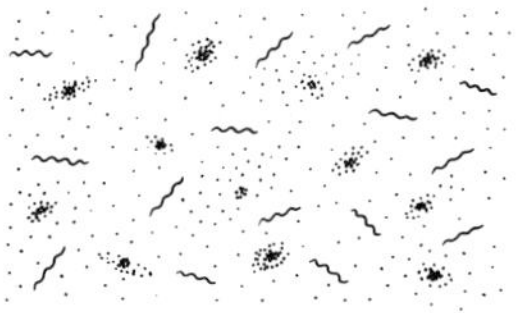	Today

Figure 7-6. Each view shows a particular portion of the universe at one of five different epochs, with a tenfold increase in the scale factor from one picture to the next. Imagine the size of each view of the region depicted as ten times larger than the previous view. The time since the big bang and the temperature of the background radiation corresponding to each view are also given. Clumps of matter, somewhat denser than the average, presumably existed in the primeval plasma. These clumps eventually collapsed to form galaxies and clusters of galaxies, while the radiation (wavy lines) propagated freely throughout the universe.

Neutrinos and Photons

If the big bang model was also valid at much earlier times than those shown in Figure 7-6, the number density (number of particles per unit volume) of neutrinos and antineutrinos should be comparable to that of photons during the epochs shown in the figure. (Since most big bang models predict an abundance of antineutrinos comparable to that of neutrinos and since their masses must be equal, from now on we shall refer to both simply as neutrinos.) The contribution of neutrinos to the present mass-energy density thus depends on their mass. If neutrinos are massless, their contribution is similar to that of photons, because neutrinos are predicted to have roughly the same energies as photons. If neutrinos are sufficiently massive, they must behave like ordinary nonrelativistic matter, because most of their mass-energy then resides in their rest mass. In this latter case, the mass-energy density of neutrinos will be proportional to their mass. If their mass equals about one hundred-millionth (10^{-8}) of the mass of a proton or neutron, the neutrinos would contribute significantly to the invisible mass that we know dominates the present universe. Recently, some laboratory experiments have hinted that at least one type of neutrino may have such a mass. In a few years, this question may be settled by more sensitive experiments.

That such a tiny mass for neutrinos would allow them to dominate the mass-energy in the universe reminds us that, in terms of *numbers* of particles, the universe probably consists primarily of photons and neutrinos. Measurements of the microwave background radiation show that about 400 million photons per cubic meter exist throughout the universe today. Recall that if we were to spread the ordinary matter (stars and gas) throughout the present universe, this dispersal would produce a uniform gas of less than one atom per cubic meter. Allowing for the possibility that the invisible matter is also composed of baryons, we find that between about 100 million and 10 billion photons exist for every baryon (neutron or proton) in the universe today. Moreover, this ratio has remained constant during the recent history of the universe, because, with only minor exceptions, both photons and baryons have been conserved (neither created nor destroyed). Because the number density of photons and neutrinos so far exceeds the number density of baryons, a correspondingly smaller mass for the neutrino would allow neutrinos to provide the same mass density as the much more massive baryons.

The Temperature of the Universe

Let us now return to Figure 7-6 to study the temperature history of the universe. We notice that the temperature T that characterizes the spectrum of the background radiation was greater in the past: The product of the universal scale factor and temperature remains constant as the universe expands. We would expect this result, because T measures the average energy (E) of the photons, and, as noted earlier, E times S remains constant.

During the first epoch shown in Figure 7-6, the high temperature prevented the nuclei and electrons from combining into atoms. As is the case inside a star, the matter at this epoch and at previous epochs had the form of a plasma: a gas composed of individual particles, rather than of atoms or molecules. Moreover, the electrons, nuclei, and photons in this plasma interacted at a rate much greater than the rate at which the universe was expanding. That is, the typical time required for a particle to undergo a reaction with one of its passing neighbors was less than the time during which the scale factor changed significantly. This circumstance guaranteed that the temperature of the matter equaled that of the radiation (which had already acquired a blackbody spectrum).

The First Cosmic Horizon

In the second epoch shown in Figure 7-6, the universe had expanded and cooled by a factor of ten. At this temperature (3,000 K) electrons and nuclei could bind together to form atoms. This change in the state of matter also produced an important change in the behavior of radiation. The photons interacted far less with the atoms than they did with the electrons in the plasma—so much less that the photons could then propagate unimpeded through space. Because the photons were no longer "coupled" to any other matter by mutual interactions (collisions), we call this epoch the "decoupling" phase. In fact, the photons freed to travel unimpeded after the formation of atoms are the same photons we observe today as the microwave background radiation. The uniform expansion of the universe preserved the blackbody shape of the photons' spectrum, characterized by a temperature that fell as the expansion of the universe lengthened the wavelengths of the photons. After this epoch, the expansion also caused the matter to cool faster than the radiation.

This epoch of decoupling, when atoms appeared, represents the first absolute horizon that limits our view of the universe. No photons can reach us directly from epochs before atoms formed, because the charged particles in the primeval plasma repeatedly scattered, absorbed, and emitted any such photons: The universe was opaque. No matter how powerful our telescopes may become, we shall never be able to "see," via any form of electromagnetic radiation, what the universe was like at such early epochs. We call this ultimate limit to our observational ability the "photon barrier." We must remember, however, that in reality it is a barrier in *time*.

Although we cannot look beyond the photon barrier, we can gain important information from accurate observations of the radiation released from the matter at the era of decoupling. Any deviations of the matter in the universe at that time from complete uniformity in *space* (inhomogeneity) will produce deviations of the microwave background radiation from uniformity in *direction* (anisotropy). If the radiation reaching us today from various directions in the sky comes from regions of the universe with properties such as density and temperature that differ from place to place, these deviations will produce corresponding differences in the intensities of the photons from different directions.

When we observe the background radiation, smaller angular sizes of the beam accepted by our telescopes correspond to smaller regions of the universe that are the source of the radiation in the beam. To date, the smallest angular size to be investigated equals about $\frac{1}{20}$ of a degree, which still corresponds to a region containing enough matter to form many galaxies. The most accurate observations at present involve larger angular sizes. We should note, however, that fluctuations on smaller angular scales than the size of the telescope beam can still produce a net fluctuation in the intensity of the radiation. We shall have to obtain observations that span a wide range of angular sizes in order to establish the extent and magnitude of any fluctuations in the properties of the primeval plasma that existed at the era of decoupling.

The Development of Structure

Apparently the universe was more uniform in the distant past, at this epoch of decoupling, than it is today, at least on the large scales we can examine. How did this apparent smoothness develop into the great diversity of structure that makes the present universe so interesting and beautiful to us? Figure 7-6 shows what we believe to be the essential nature of this process of evolution—the same process, we believe, that has also led to the formation of stars from interstellar gas and dust in galaxies. This process is *gravitational collapse.*

Gravitational collapse cannot occur unless the seeds of growth already exist. These "seeds" are regions with somewhat higher density than their surroundings, such as those indicated by the darker shading in the first two epochs shown in Figure 7-6. If a region of higher density has sufficient size, the inward pull of gravity on the matter in that region overwhelms the opposing tendency of pressure forces to disperse the matter toward uniformity. The matter in that region will then collapse. This region of slightly higher density is initially expanding, as are its surroundings. However, as is the case in a closed universe, the gravitational force can eventually convert the expansion into a contraction if the initial density enhancement is sufficiently large. (Recall that we must mentally enlarge each succeeding picture in Figure 7-6 by a factor of ten. During the first three epochs the actual size of the clumps still increases, even though they are contracting relative to the surrounding matter.)

Although gravitational collapse seems to play an essential role in the evolution of large astronomical objects, other aspects of structural development are far less certain. The most fundamental uncertainty concerns the origin of the fluctuations in density. In the standard big bang scenario, these seeds of growth are primordial, in existence well before the epochs depicted in Figure 7-6.

On the one hand, if these density fluctuations involved both the ordinary matter (charged particles) and radiation, the fluctuations would survive as sound waves provided that the "frictional" forces from the interaction of the matter and radiation gases were strong enough to prevent their relative motion. This would be possible for fluctuations that contained at least the mass of several galaxies. Such fluctuations would then persist until the era of decoupling, at which point they would collapse. On the other hand, if only the matter was involved in the fluctuations, all sizes of fluctuations would remain "frozen" in the photon gas. Here, the same sort of photon friction would prevent the matter fluctuations from either growing or weakening before the era of decoupling. After decoupling, the pressure of the matter would prevent the collapse of these fluctuations only if the mass contained in a fluctuation was less than about 100,000 times the mass of the sun. Although these two possibilities lead to definite, and different, predictions for the range of masses that can initially form, this analysis defers the question of the origin of the fluctuations to an earlier epoch.

These two possibilities for the nature of the density fluctuations—matter plus radiation or matter alone—lead to different scenarios for the creation of the wide range of masses that we observe today. The first scenario, called the *fragmentation hypothesis,* begins with the extremely large-scale fluctuations that result from the situation in which matter and radiation fluctuate together. During the process of collapse, these initial clumps would fragment into subunits of decreasing mass, eventually of galactic size, with the galactic-size clumps later fragmenting into stars.

By contrast, the second scenario, called the *clustering hypothesis,* begins with objects of the minimum mass allowed by the second possibility, in which the matter alone fluctuates in density. These minimum-mass clumps presumably would collapse before any objects of higher mass could do so because these clumps initially would have had a greater density enhancement. Whatever the final fate of these clumps, which may have become supermassive stars, the gravitational forces between them would tend to make them cluster together, first into objects of galactic size, and then into the clusters of galaxies that we observe today. Much smaller objects (ordinary stars) would presumably have formed from the fragmentation of these supermassive stars and from the leftover gas, with some supermassive stars possibly surviving.

A related scenario, called the *explosive hypothesis,* would generate large-scale fluctuations in density from the shock waves that propagate outward into the surrounding medium after the explosion of the supermassive stars. These large-scale fluctuations would then collapse to form galaxies and clusters of galaxies.

These scenarios become more complicated if neutrinos have a mass of the order required to provide the density of invisible matter. As the neutrinos' energies decreased because of the expansion of the universe, their velocities must then have eventually fallen far below the velocity of light. This reduction of their velocities would allow the neutrinos to clump together by gravitational forces, along with the ordinary matter. The details of this clumping process would depend on the neutrinos' mass and are not yet completely understood. It is clear, however, that the only fluctuations in the density of neutrinos that can grow initially are those that encompass a region containing enough ordinary matter to form at least a thousand galaxies.

We also face many unanswered questions concerning the most recent phases of galactic evolution (the last two epochs shown in Figure 7-6). What factors determine whether a fluctuation will develop into a spiral, an elliptical, or some other form of galaxy? Did most galaxies develop at their center a supermassive black hole, or some other efficient source of the huge amounts of energy that we have observed emanating from them? Was this source most active during the adolescence of a galaxy? If so, this activity would explain the greater number of quasars and other violent galaxies compared with the number of ordinary galaxies as we observe larger red shifts and hence look further back in time.

The Age of the Universe

Although there are many uncertainties, this general picture of recent cosmic evolution is at least consistent with all our knowledge of our cosmic environment; it provides a framework upon which we can develop a still deeper understanding of the cosmos. An important example of this consistency is the investigation of the age of the universe.

The universe must have existed at least as long as it has been expanding. We have determined that this "expansion age" is at most 10 to 20 billion years, depending on the exact value of the Hubble constant, the present rate of expansion. The deceleration of the universe makes the actual expansion age less than this limit, but observational restrictions on the size of the deceleration parameter show that this difference is not large. The possible range of expansion ages (for a given value of H_o) is shown in Figure 7-4 by the range of times in the past (corresponding to different values of the deceleration) when the distances between particles were close to zero.

From our scenarios of how structure in the universe developed, we conclude that all stars must be younger than the expansion age. The 12 to 18 billion years calculated for the ages of the oldest stars agrees with this requirement, if the Hubble constant and the deceleration parameter are not too large. A final check on consistency comes from the age determinations of those heavy elements that decay radioactively into other elements. By knowing the decay rates and the present abundances of a variety of such elements, we can calculate the time since the decaying elements formed. This independent method gives ages for these elements in the same range as the ages of the oldest stars. We expect this result, since we believe that all such (heavy) elements were synthesized within stars.

Was the Early Universe Hot?

An intriguing relationship exists between the ordinary matter and the radiation content of the universe, the possible significance of which was first realized by the British astrophysicist Fred Hoyle in 1965, soon after the microwave background radiation was discovered. Calculations had shown that the nuclear reactions within a star convert about 1 percent of the star's rest-mass energy into radiation (starlight) during the star's lifetime. One might therefore expect the mass-energy density of starlight in the universe today to equal about 1 percent of the mass density of stars that have completed most of their evolution. And indeed such a relation exists between the amount of starlight and the number of such stars.

What struck Hoyle was the fact that the mass-energy density of starlight within galaxies equals that of the microwave background radiation. Is this equality just a coincidence, or does it have profound cosmological implications? Could the microwave background radiation have emanated from stars, and not from a hot gas that filled the early universe, as we have assumed? Did starlight become microwave photons? If so, was the early universe much colder than we have assumed?

Each starlight photon has an energy roughly 1,000 times greater, and a wavelength 1,000 times less, than the corresponding values for a microwave photon. In order to convert a significant fraction of starlight photons into microwave photons, sufficient matter must be dispersed throughout space (in the form of interstellar dust, for instance) to absorb the starlight photons and emit their energy in the form of microwave photons. Moreover, both the stars and the matter must have a fairly uniform distribution in space in order to produce both the precise blackbody spectrum and the observed isotropy of the microwave background. We know enough about the contents of the present universe to conclude that this conversion of starlight photons into microwave photons cannot occur today.

But could this conversion have occurred in the past? Suppose, for instance, that an early generation of stars formed sometime between the second and third epochs in Figure 7-6, but in a universe that was initially much cooler than we have assumed in our standard big bang model. If these stars were much more massive than the sun, they would have shined for only a brief time (less than 10 million years) before they collapsed into dark remnants (black holes or neutron stars). They would therefore not be visible today, but could provide the invisible mass that dominates the universe. In addition to their now-vanished starlight, such stars might have efficiently generated other forms of radiation, for example, X rays (by the accretion of gas onto their remnants, as discussed in Chapter 5). If these stars expelled a sufficient fraction of their mass after they had produced some heavy elements in their interiors, successive generations of such stars could have produced the interstellar dust grains required to absorb and reemit the radiation of existing stars.

If these massive stars also produced and expelled a large amount of helium, this process could account for the large difference between the observed abundance of helium in the universe today and the amount that has been produced by the stars we see now. But this missing helium also emerges naturally in a hot early universe, as we shall see in Chapter 8.

At this point, the astute reader may be troubled by the expansion of the universe, for, as noted, the expansion requires that the microwave background photons had greater energy in the past. The amount of radiation produced by this hypothetical early generation of stars in the "cold" big bang model would therefore have had to be correspondingly greater than that produced by a present generation of such stars, because of the subsequent red shifting of the radiation. This energy requirement in turn requires that more mass resided in stars at that early time. However, this scenario is marvelously consistent, because the amount of invisible mass (the remnants of this early generation) must exceed the visible mass (stars) in the present universe.

Certain features of this cold big bang model are appealing to some physicists and astronomers. In particular, the model explains in a natural way why a few billion photons per baryon exist in the universe today. However, the properties of a cold early universe are not at all clear, and the hot big bang model can explain the photon-to-baryon ratio in another way (see Chapter 8). Still, we cannot now rule out the cold model.

In the next chapter, we shall explore the consequences of the standard (hot) big bang model as we extend it further back in time, since this model is the simplest and most straightforward extrapolation of our knowledge of the present to earlier epochs. However, the alternative model of a cold early universe illustrates the need to explore all such possibilities until observational evidence, the ultimate criterion, allows us to determine what the early universe was really like.

"I think you should be more explicit here in step two."

8

Beyond the Photon Barrier

For violent fires soon burn out themselves
William Shakespeare, *King Richard II*

WE NOW EXTEND OUR INVESTIGATION of the history of matter further back in time, beyond the photon barrier to what we shall call the early universe. (This era corresponds to times less than 100,000 years after the expansion began.) In doing so, however, we must recognize the risks as well as the rewards. The risks arise from the increasing degree of uncertainty in our knowledge of the forms of matter and the laws that govern matter as we extend our frontier into earlier epochs. As we probe deeper into the past we find a universe filled with particles of higher energies. At the earliest times, these energies exceeded any that have been, or will ever be, achieved in any man-made particle accelerator. Hence we have fewer experimental checks on the theories that must guide us. Moreover, we shall never be able to view (via photons) the matter that existed at these early times. Nonetheless, the rewards of probing these early epochs are great. This investigation holds the promise of yielding a deeper understanding of the origin of everything we see around us today.

The Cosmic Cauldron

Before we describe what the early universe was like according to the standard (hot) big bang model, let us briefly review the basis for this description. Einstein's universal evolution equation tells us how rapidly the universe changes over time. The equation becomes especially simple when applied to the early universe, when the scale factor was much less than one. As we can see from Figure 7-3, the magnitude of the curvature term of the equation was negligible compared with the magnitude of the density term when S was extremely small. Therefore the square of the

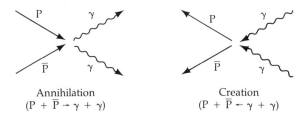

Annihilation
$(P + \bar{P} \rightarrow \gamma + \gamma)$

Creation
$(P + \bar{P} \leftarrow \gamma + \gamma)$

Figure 8-1. When a particle (P) and an antiparticle ($\bar{P}$) collide, they can annihilate, converting their energy into the energy of lighter particles, such as the photons shown here. The reverse process, creation of a particle and its antiparticle, is also shown. It will occur if the lighter particles (in this case photons) have sufficient energy to supply at least the rest-mass energy of the resulting heavier particle and antiparticle.

expansion rate was then simply proportional to the total mass-energy density. The higher the density, the more rapidly was the universe expanding. But the density of particles decreased as the matter expanded. Hence the further back in time we look, the higher was the density of particles—but the more rapidly was that density decreasing because of the faster universal expansion.

We have already discussed in Chapter 7 how the expansion of the universe reduces the energy of relativistic particles, such as photons. Non-relativistic particles also lose kinetic energy as the universe expands, just as an ordinary gas cools during expansion. But since most of the total energy of such ordinary, nonrelativistic particles consists of their rest-mass energy (which is fixed, unlike their kinetic energy), this energy loss is not significant.

We have also seen that as the universe evolves, any type of particle remains in thermal equilibrium with the other constituents of the gas (and therefore at the same temperature) so long as it interacts with them through collisions. The temperature provides a measure of the particles' kinetic energies.

Pair Creation and Annihilation

We must now consider an important process that occurs at the extreme energies that characterized the early universe: the production and annihilation of particle-antiparticle pairs. Figure 8-1 shows the annihilation of a particle (P) and its antiparticle ($\bar{P}$). The particle and antiparticle can disappear if their separation becomes sufficiently small, in which case a less massive particle and its antiparticle emerge in their place. In Figure 8-1 the emerging particles are photons, which are their own antiparticles. Figure 8-1 also depicts the reverse process, pair creation, in which a particle and its antiparticle are produced by the collision of a lighter particle and its antiparticle (in this case, two photons). This process cannot occur unless the lighter pair of particles has sufficient energy to supply at least the rest-mass energy of the heavier pair. If any particle's average kinetic energy (corresponding to a given temperature) plus its rest-mass energy exceeds the rest-mass energy of another type of particle, that other type of particle and its antiparticle can be created (assuming that the particular creation process is allowed by the laws of physics). This condition was satisfied at extremely early times in the universe, when the temperature was high enough to give the light particles the energy required for pair creation.

Moreover, at those times when this energy condition was satisfied, the density of photons, neutrinos, and other light particles was so high that the rate at which pairs of new particles and antiparticles appeared exceeded the rate at which the universe was changing because of its expansion. This fact requires that the rate of the reverse reaction—annihilation— must have been correspondingly great, in order for the annihilations to balance (approximately) the creations. We shall now see how the resulting abundances of particles and their antiparticles evolved.

Particle Democracy and Particle Survival

At temperatures corresponding to kinetic energies much greater than its rest-mass energy, each type of particle and antiparticle had roughly the same abundance as the photons in the primeval plasma. This is not surprising, since such particles were then relativistic and therefore behaved like photons. As the universe expanded and the temperature dropped below the value at which the kinetic energy of each particular type of particle was comparable to its rest-mass energy, the abundance of that particle type began to decrease rapidly with respect to the photons, because fewer and fewer particles had energies sufficient to create that type of particle. When the density and temperature of the particles had decreased to the point at which these creation and annihilation rates became less than the expansion rate of the universe, the reactions effectively ceased. At this point, cosmologists say the particular type of particle had been "frozen out" of the cosmic sea: Its abundance thereafter remained essentially fixed. The larger the rest-mass energy of any particular type of particle, relative to the temperature at which this freeze-out occurred, the smaller the number of particle-antiparticle pairs that remained, unable (because of their low abundance) to find and to annihilate one another. The unstable particles later decayed into other kinds of particles.

This scenario describes what happened to those types of particles that had the same abundance as their antiparticles. If a sufficient excess of particles over antiparticles (or vice versa) existed, then the final abundance of particles (or antiparticles) equaled that excess, and the final abundance of antiparticles (or particles) was far less than this value. We can see that the reactions considered here changed only the *sum* of the number of particles and antiparticles, not the *difference*. Hence the difference survived with whatever value it had originally.

For instance, we have discussed why we believe that at least our local region of the universe contains almost no antinuclei, and therefore has a net (positive) baryon number. We have even stronger evidence that the net electrical charge of any region of the universe must be zero. Therefore the positive charge of the nuclei must be balanced by an equal negative charge; hence the number of electrons must far exceed the number of positrons. Although many antineutrons, antiprotons, and positrons existed in the early universe, virtually none of them has survived. These antiparticles were annihilated by the higher abundance of neutrons, protons, and electrons.

A History of the Early Universe

The concepts we have discussed here and in the previous chapter provide the major components of the simplest big bang theory, the standard (hot) model, which allows us to reconstruct a possible history of the universe. Figure 8-2 shows this history by depicting the evolution of the mass-energy density of the various particle species that existed. It also shows how the scale factor and the average energy of the relativistic particles (which dominated the early universe) have varied with time. For comparison, the rest-mass energies of a proton ($m_p c^2$) and an electron ($m_e c^2$) are marked on the energy scale. Time is measured from the big bang, the epoch of virtually infinite density and temperature (extreme left of Figure 8-2) from which the universe expanded.

Let us now work our way back through time on the graph to uncover our early history. In the previous chapter we looked backward through the more certain "recent" past, up to the photon barrier shown in Figure 8-2. From this graph, we see again that if neutrinos (v) are massless, most of the mass-energy in the present universe resides in the form of nuclei, unless some new type of particle (such as magnetic monopoles, discussed later) fills the universe. The alternative neutrino curve in Figure 8-2 (labeled $m > 0$) corresponds to a neutrino mass sufficient to make the mass-energy density of neutrinos equal to the critical density, thereby producing a flat universe (one with zero curvature). Any larger mass for the neutrino would produce a closed universe (one with positive curvature).

As we have discussed, once the expansion of the universe had reduced the velocities of electrons, nuclei, and any other particles with mass to values well below the speed of light, the mass-energy density of these particles decreased less rapidly as the universe expanded than did the mass-energy density of photons and any other massless particles. Note from Figure 8-2 that any massive neutrinos would have had such nonrel-

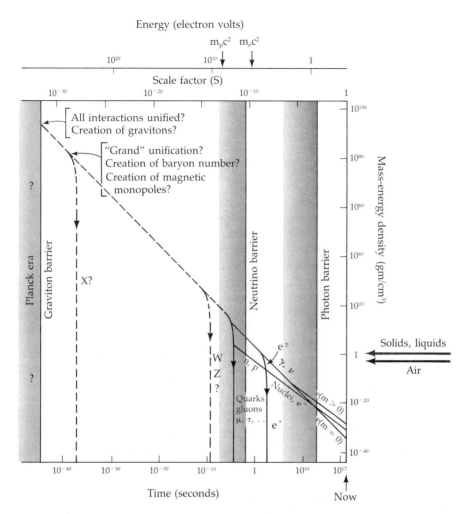

Figure 8-2. The history of the universe is depicted here in terms of the mass-energy density of the different types of particles that were present at various times since the big bang. For comparison, the densities of solids, liquids, and air are marked at the right. The evolution of the scale factor (S) and the average energy of relativistic particles are also indicated. For comparison, the rest-mass energies of a proton ($m_p c^2$) and an electron ($m_e c^2$) are marked. At early times, many types of elementary particles (X, W, Z, quarks, gluons, μ, τ, e, γ, ν, . . .) existed. Most disappeared because of particle-antiparticle annihilation when the particles' kinetic energies became less than their rest-mass energy. Neutrons (n) and protons (p) were produced from quarks at about 10^{-5} seconds, and nuclei were produced from nucleons at about 10^2 seconds. The three barriers shown indicate the epoch beyond which we cannot "see" via each of the three types of particles. In addition, at energies higher than that of the "accelerator barrier" (indicated by the dashed portion of the curves), we have little direct knowledge of the laws of physics.

ativistic speeds, and therefore would have behaved unlike massless neutrinos, at times greater than about a year (3×10^7 seconds) after the big bang.

The Transition from Nucleons to Nuclei

Proceeding back through time, we find that the first era of particular interest beyond the photon barrier spans the period of time from about $\frac{1}{100}$ of a second to a few hundred seconds after the big bang. During this era the kinetic energy of a typical particle was comparable to the rest-mass energy of an electron. Three events of interest occurred during this era. First, the neutrinos ceased to interact with the other particles, just as the photons ceased to interact a million years later. We may therefore call this epoch the "neutrino barrier" (see Figure 8-2). Second, the abundance of electrons and positrons began to fall below that of photons and neutrinos because of annihilation, which eventually left a residue of electrons. Third, certain nuclei formed from the nucleons (neutrons and protons) then present. We call this entire era the nucleon-to-nuclei transition era, which we shall describe more fully later in this chapter.

The Transition from Quarks to Nucleons

Further back in time, the next era of interest, which we shall call the quark-to-nucleon transition era, began only a millionth (10^{-6}) of a second or so after the big bang, when particles had kinetic energies close to the rest-mass energy of a neutron or proton. During this time, the composition of the universe changed even more drastically than during the subsequent nucleon-to-nuclei transition era. Just before the quark-to-nucleon transition era began, the universe consisted of a gas of elementary particles, most of which moved at speeds at or close to the speed of light. The particles included photons, electrons, electron neutrinos, heavier leptons (μ and τ) and their associated neutrinos, quarks, and gluons, plus an almost equal abundance of all the antiparticles of these particles. Any other type of particle (not yet discovered) whose mass was small enough that it had not yet suffered net annihilation must also have existed with an abundance close to that of the other particles. Since the heavier leptons have rest-mass energies comparable to those of a neutron or proton, they began their net annihilation at about that energy, and the survivors eventually decayed. Mesons formed from quarks and antiquarks, but they also soon decayed. The excess of quarks over antiquarks survived the quark-to-nucleon era in the bound systems of quarks: neutrons and protons. No free quarks or gluons remained, according to the prevailing theory (quantum chromodynamics) that governs them.

The Accelerator Barrier

As we move still further back in time to higher energies, greater than about ten times the rest-mass energy of a proton, we encounter another barrier to our understanding. This limit, which we shall call the "accelerator barrier," exists because our present accelerators cannot reach collision energies as high as those that existed at such early times, less than a billionth (10^{-9}) of a second after the big bang. We have only a few other types of experiments (such as the detection of magnetic monopoles or of decaying protons), which provide less direct information, to guide us in our choice of the laws of physics at such high energies. For this reason, the portion of Figure 8-2 beyond the accelerator barrier consists of dashed lines, in order to reflect the much greater uncertainty in our theory of the universe then. The accelerator barrier to our knowledge differs from the photon barrier and the neutrino barrier in that the former is a limit imposed by our technology (and will be pushed back to somewhat higher energies in the future), whereas the latter mark fundamental limits on any civilization's ability to investigate the universe.

At energies only a little beyond the accelerator barrier (about a hundred times the rest-mass energy of a proton), we should find the particles that transmit the weak interaction—the W boson and its antiparticle, as well as the uncharged Z boson. Like the heavier leptons, these particles are unstable, but that fact did not prevent them from acquiring the same abundances as the other particles so long as they were relativistic, since they were created (and annihilated) much more rapidly than they decayed. In a few years we shall have accelerators energetic enough to produce these heavy particles, if they exist. Persuasive experimental and theoretical evidence leads physicists to believe that the W and Z bosons do exist; if so, their discovery will confirm our belief in the unification of the weak and electromagnetic interactions (see Chapter 3). This unification means that at energies higher than the rest-mass energies of these particles, the differences between the electromagnetic and weak forces, so apparent in our present low-energy world, disappear. We therefore conclude from Figure 8-2 that at times earlier than 10^{-11} seconds after the big bang, only three distinct fundamental interactions existed: gravitational, strong (color), and electroweak.

Higgs Bosons

At such early times, a quite different type of particle may also have existed: what we call "Higgs bosons," named in honor of the physicist Peter Higgs (who demonstrated in 1964 why they might exist). Higgs bosons were "invented" to explain certain properties of other particles. The interaction of Higgs bosons with the carriers of the weak interaction, the

W and Z bosons, provides a natural way to explain how the latter bosons acquired a large mass, whereas the carriers of the electromagnetic and gravitational interactions, the photon and graviton, remained massless. In giving mass to the W and Z bosons, Higgs bosons themselves acquired a mass of the same order of magnitude, too great to have been produced in accelerators thus far.

While inducing masses in other particles, Higgs bosons may have done something still stranger. They may have *reduced* the mass-energy density of empty space (no particles present). We believe that empty space contains little or no mass-energy density today; otherwise we would have seen its gravitational effects on the expansion of the universe. Therefore empty space might have had a large, positive mass-energy density in the early universe. This additional density could have greatly increased the expansion rate (recall the evolution equation). Its contribution would have fallen to zero when the Higgs bosons acquired their mass. In effect, the mass-energy of empty space would have been converted into additional mass-energy of Higgs bosons and other particles. The Higgs bosons, being unstable, eventually decayed.

The theoretical properties of Higgs bosons and their cosmological consequences are now under active investigation. However, for the most part we shall continue to base our description on the simplest big bang model (which does not allow for any such contribution to the mass-energy density of empty space). This description may have to be modified if Higgs bosons are discovered in more powerful accelerators.

Grand Unified Theories

During the 1970s, following the apparent confirmation of the electroweak unification theory, many physicists came to believe that the strong interaction also participates in the unification of forces, but at a much higher energy. The value of this unification energy, above which only two distinct interactions (gravitation and grand unified) exist, remains uncertain but is incredibly large. Figure 8-2 indicates a possible value (roughly 10^{15} times the rest-mass energy of a proton, corresponding to incredibly short times, roughly 10^{-38} seconds after the big bang) and also depicts the evolution of the proposed extremely massive bosons (called X) associated with this grand unification. Of course, this unification will not be truly "grand" until gravitation is included, fulfilling Einstein's belief that all forces in nature in reality must be one.

A grand unified theory potentially has profound importance for cosmology, not least because it could explain the apparent preponderance of matter over antimatter in the present universe. Note first that unification of the strong force that differentiates between quarks and leptons implies that at high energies many differences between quarks and leptons disappear; in fact a quark can become a lepton (through the exchange of an X boson) and vice versa. If this transformation is possible, then baryons can be transformed into leptons or can be created from leptons. Hence neither the total baryon number nor the total lepton number of the universe would be conserved. We may be able to observe this breakdown of baryon conservation in the present universe by searching for the predicted, though extremely rare, decays of protons and other nuclei. Since we know from experiment that the lifetime of a proton exceeds 10^{30} years, many tons of material (such as water) must be observed continuously to detect even a single proton decay within a reasonable period of time. If we find such a decay, we shall have the first evidence of the instability of "ordinary" matter.

The decay of protons would provide strong evidence that some theory of grand unification must be valid. Then we might reasonably entertain the aesthetically appealing belief that in the incredibly distant past, the universe had complete symmetry with respect to matter and antimatter. Only when the particle energies fell to the point at which the effects of the grand unification (such as the X bosons) began to disappear could the net baryon number (baryons minus antibaryons) begin to deviate from zero. However, this change also requires that the laws of physics be asymmetric with respect to particles and antiparticles, so that the number of antibaryons (antiquarks) could decrease relative to the number of baryons (quarks). This would have left the universe with its present excess of baryons over antibaryons. (By contrast, if the symmetry between baryons and antibaryons was not "broken," the annihilation of the equal numbers of baryons and antibaryons would have left the present universe with a billion times fewer nuclei per photon than the value we observe, plus a number of antinuclei equal to that of the nuclei.) This scenario for the production of a matter-antimatter imbalance was first proposed by the Russian physicist Andrei Sakharov, winner of the Nobel Peace Prize in 1975.

One aspect of this proposal for the origin of matter seems unsatisfying at present. The type of asymmetry in the laws of physics that is required to produce more baryons than antibaryons has been detected (by Val Fitch and James Cronin, for which they were awarded the 1980 Nobel Prize in physics). But the process discovered by Fitch and Cronin does not at all resemble the type of process required in the early universe. Hence the degree of asymmetry, which determines the magnitude of the baryon excess produced, remains an unknown quantity. Any such theory of the baryon excess produced within some grand unified model can therefore *explain* the number of baryons per photon in our universe, but does not uniquely *prescribe* this ratio. This looseness contrasts with the rival theory involving a cold early universe, described at the end of Chapter 7, which produces a range of/possible values of the ratio of baryons to photons that more tightly brackets the observed value.

Another possible consequence of grand unification arises from the fact that additional Higgs bosons (which give the X bosons their masses) are usually predicted to exist at this early epoch. Their existence seems to lead to the production of *magnetic monopoles*, stable particles with magnetic, rather than electric, charge. Like ordinary electrically charged particles, magnetic monopoles produce electromagnetic fields, but with the roles of the electric and magnetic fields reversed. That is, a fixed magnetic monopole produces a magnetic field, whereas a moving magnetic monopole produces an electric field in addition.

Magnetic monopoles have never been detected with certainty. Moreover, most grand unified theories predict that they should have extremely large masses, roughly the same as those of the X bosons and associated Higgs bosons. This huge mass (about 10^{16} times greater than that of a proton), combined with the theoretical estimates of the initial abundance of magnetic monopoles, would lead to a huge mass density of magnetic monopoles today. In fact, the mass density of magnetic monopoles would be so large that we would not be here—the universe would have collapsed long ago. Hence any theory of grand unification must somehow avoid this overproduction of magnetic monopoles.

However, such constraints on the laws of physics presuppose the validity of the standard big bang model. As we have discussed, this validity must be considered extremely uncertain for the early epochs involved in grand unification.

The Final Barrier

As we move still further back in time beyond this epoch of grand unification and baryon production, we reach the final barrier shown in Figure 8-2. Here we encounter the Planck era, named in honor of physicist Max Planck, who in 1899, before Einstein had formulated the special or the general theory of relativity, first proposed the discrete (quantum) nature of energy, which can drastically alter the properties of gravitation under the extreme conditions that may have existed at such early moments of the universe. Like the other fields that describe the fundamental interactions between particles, the gravitational field should be quantized: It should be describable as composed of individual particles (gravitons), just as the electromagnetic field can be described in terms of photons. Since gravitation affects all matter, it determines the properties of space and time. Hence the properties of space and time (that is, the geometry of the universe) must therefore also be quantized. This quantization in turn results in fluctuations of the gravitational field, so that the geometry at every point in space and time should be rapidly changing (in discrete "jumps" rather than smoothly).

The effects of this profound alteration in our conception of gravitation appear only at the incredibly high energies (and densities) that characterize the Planck era shown in Figure 8-2. At the present time, we have no theory to guide us at times earlier than the "Planck time," which marks the end of the Planck era. (The corresponding mass and size of the visible universe is represented by the Planck point in Figure 5-5). Since gravitons can propagate to us undisturbed at later times, we can also call this time the "graviton barrier" (as shown in Figure 8-2). The graviton barrier therefore plays the same role as the photon and neutrino barriers.

Whereas the accelerator barrier represents a present limit of our experimental ability, the graviton barrier represents a present limit of our theoretical ability. Any discussion of what the universe might have been like before the Planck time, during the Planck era, lies outside the present realm of science. We can hope that someday the final unification of gravitation with the other types of forces will tell us how the universe behaved during the Planck era. In such a completely unified theory, only one type of interaction would have existed between particles in the universe during the Planck era. Only later, as the universe expanded and cooled, would the forces that we know today have assumed their separate identities and strengths.

The Daring Extrapolations of Cosmology

Having now come to the end of our story of the history of the universe, let us reflect for a moment on this daring extrapolation into the past. Note from Figure 8-2 that the mass-energy density in the universe at the Planck time was 123 orders of magnitude (10^{123} times) greater than it is today, corresponding to the fact that the average particle separation was 32 orders of magnitude smaller then. (For comparison, Figure 8-2 also shows the relatively minuscule density of ordinary solids and liquids and that of the air we breathe.) In addition, the particle energies at the Planck time were 17 orders of magnitude greater than those we can achieve in our accelerators, and 14 orders of magnitude greater than the effective collision energies of the highest-energy cosmic rays. Finally, the time during which the universe expanded by a factor of two (approximately equal to the inverse of the expansion rate) was 61 orders of magnitude less at the Planck time than it is today.

How can we have any faith in our conclusions, given the incredible extrapolation from normal physical conditions that we must attempt in order to consider the earliest moments of the universe? Can we check, in any way, whether the early universe really was the hot gas of particles we have described? Fortunately we can, since some relics of the early universe may persist throughout our universe.

Relics of the Early Universe

Although the photon barrier obstructs our vision, we can nevertheless obtain information from earlier epochs. This information reaches us in two forms, both of which are physical remnants of the early universe. The first type of relic consists of those particles, produced in the early moments of the universal expansion, which have lifetimes (against decay) greater than the expansion age of the universe, so that they still exist today. The second type of relic is the large-scale distribution of matter (such as galaxies and clusters of galaxies) described in Chapter 6. This distribution appears to have arisen from tiny fluctuations in the otherwise uniform gas that filled the early universe. One of the major unsolved problems in cosmology is the origin of these fluctuations. Cosmologists have proposed other scenarios, in which a universe extremely chaotic in its earliest moments was subsequently "smoothed out" by a variety of proposed processes. Whatever the origin of the fluctuations, the distribution of matter today may carry imprints of its birth long ago, clues that we hope to use to learn about conditions in the early universe.

Such methods of studying long-vanished eras are cosmic archaeology: sifting the present to find ashes of the past. We shall discuss in some detail what may be the most revealing relic of the early universe that we possess today—the abundance of certain types of light nuclei. Before we do this, however, let us consider some of the other types of stable particles that might have been born in the primeval plasma.

Particles as Probes of the Early Universe

We have already discussed the neutrinos, which should still be as abundant as photons but are extremely difficult to detect. Even harder to detect are the gravitons, which may have been produced during the Planck era with an abundance also comparable to that of photons. Gravitons would have been produced at later times as well in the form of the gravitational radiation generated by the collapse or interaction of massive bodies. Magnetic monopoles, if they are discovered, would likewise provide an important clue to conditions in the early universe.

Since 1981 a group of experimental physicists at Stanford University headed by William Fairbank has claimed to have detected particles with electric charges equal to $\frac{1}{3}$ or $\frac{2}{3}$ the charge of a proton or an electron. (Recall that quarks have precisely these fractional values of charge.) However, all other groups searching for fractionally charged particles have been unsuccessful. The currently accepted theory of strong interactions, quantum chromodynamics, predicts that the force that binds quarks together to form mesons and nucleons has so much strength that quarks can never be separated as isolated, free particles. Hence if Fairbank's observation of fractional charges is confirmed by other experiments, we shall have to change our current beliefs, either by modifying the theory of quantum chromodynamics or by introducing a new type of particle.

If fractionally charged particles exist, they may furnish us with another important probe of the early universe. Their present abundance could reflect the physical conditions at the time they were "frozen out" of equilibrium. But as with other possible relic particles, we must remember that fractionally charged particles might have been produced in other ways—for example, by collisions of high-energy cosmic rays with nuclei near the surface of the earth.

Our examples underscore the important fact that *any* type of stable free particle, even one so massive that it cannot be produced on earth, can exist today as a relic of the past history of matter in the universe. The prediction of the present abundance of such particles is playing an increasingly important role both in cosmology and in elementary particle physics.

Nuclei as Probes of the Early Universe

Let us now consider in more detail the key relic that we mentioned earlier: nuclei. Let us go back in time to about $\frac{1}{100}$ of a second after the big bang, when a typical particle had a kinetic energy ten times the rest-mass energy of an electron. From Figure 8-2, we can see that the particles that filled the universe at that time were photons, neutrinos, antineutrinos, electrons, and positrons (all nearly equal in abundance), as well as a much smaller number of neutrons and protons. Bear in mind that collisions among these particles, often involving conversions of one particle type into another, proceeded more rapidly than the expansion of the universe at this time. That is, a typical particle interacted many times with other particles during the time needed for the large-scale properties of the universe (such as density and temperature) to change significantly. As a result of the frequent collisions, interactions among particles established "statistical equilibrium," a condition in which basic properties of the primeval plasma, such as the abundance of most particle species, depended only on the temperature at that time. Since we know the evolutionary history of the temperature from the equations of our big bang model, we therefore also know most of the characteristics of the matter at any time during the expansion of the universe.

It is important to note that these particle abundances did not depend upon the earlier history of the universe, but rather depended only on the conditions at the time of interest. The many reactions among particles erased most of the information they could have carried from earlier epochs. Some features of the universe persisted, however, since the laws of physics demanded it. Two quantities that remained constant since well before this era when nuclei were created were the net number of baryons (baryons minus antibaryons) and the net number of leptons (leptons minus antileptons.)

The relevant measure of the number of baryons—in this era protons and neutrons—was the number of baryons per photon. This ratio, which we have seen has an extremely small value in the present universe, decreased only slightly from this early era to the present, as positrons annihilated with the electrons to produce more photons. The present baryon-to-photon ratio is proportional to the present mass density of ordinary matter, since most of the mass of the atoms that constitute this matter resides in their nuclei (composed of protons and neutrons). We know the present density of photons quite accurately from the observed temperature of the microwave background radiation. But we do not know the average density of ordinary matter, and thus the density of baryons, with as much accuracy. Hence the present baryon mass density must remain an adjustable parameter in our big bang model.

The other conserved quantity, the lepton number, equals the number of electrons and neutrinos minus the number of positrons and antineutrinos, since the heavier leptons had decayed into other particles by this time. In the standard big bang model we usually assume that within any volume containing many particles, the lepton number—like the baryon number—must be much less than the photon number. We shall see later what happens if we discard this assumption as well as the other assumptions from which we construct our big bang model.

Proceeding with our description of the evolution of the universe during this era, we find that a major change occurred about one second after the big bang. At that time, the rate at which particles interacted through the weak force became less than the expansion rate of the universe. This change had two major consequences. The first, as previously mentioned, was that the neutrinos could then propagate freely throughout the universe, basically oblivious to the presence of any other particles. Even more important, the weak reactions responsible for the conversion of neutrons into protons and vice versa—reactions 1a, 1b, 2a, and 2b in Figure 8-3—effectively ceased. This "froze" the ratio of neutrons to protons at approximately the value it had at that time.

Ordinarily, neutrons and protons in a gas with a density such as existed at this time would fuse during collisions to form deuterium (see reaction 3a in Figure 8-3). However, at temperatures as high as those that then existed, photons had enough energy to destroy deuterium nuclei (see reaction 3b in Figure 8-3). Not until the temperature dropped enough to reduce this destruction rate significantly could the abundance of deuterium grow to an appreciable value.

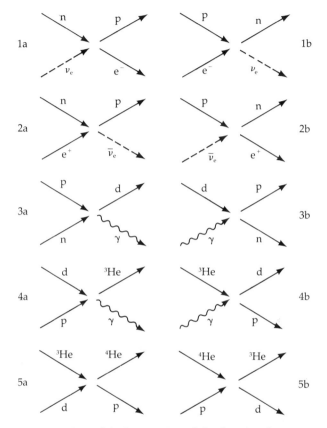

Figure 8-3. Here is a sampling of the key reactions (left column) and reverse reactions (right column) involved in the production of nuclei in the early universe. "Weak" reactions involving electrons (e⁻), neutrinos (ν_e), and their antiparticles (e⁺, $\bar{\nu}_e$), convert neutrons (n) into protons (p) (1a, 2a) and protons into neutrons (1b, 2b). Other key reactions are the production (3a) and destruction (3b) of deuterium (d), which involves photons (γ); and the production of ³He (4a) and ⁴He (5a).

Once deuterium became abundant, a few hundred seconds after the big bang, a host of other reactions became possible. Two of the chief reactions are drawn as 4a and 5a in Figure 8-3. By this time, most particles had energies insufficient to allow the reverse reactions (such as 4b and 5b, in which the heavier nuclei are destroyed) to occur. Thus the buildup of heavier nuclei (e.g., ^{3}He and ^{4}He) could proceed. The low values of the density of colliding nuclei and of the time available for collisions—in comparison with the values within stars—meant that only the lightest of these heavier nuclei could be produced in significant numbers. As the expansion of the universe reduced its temperature and density, nuclear reactions effectively ceased until stars formed much later.

Fortunately, the rates of all these reactions have been measured in accelerator experiments. We therefore have confidence in our understanding of the physical processes that occurred during this era, a few hundred seconds after the big bang. By contrast, for epochs preceding the accelerator barrier, we have primarily theory, rather than experiment, to guide us.

Calculated and Observed Abundances of Nuclei

The abundances of the various nuclei that emerged from this era are shown in Figure 8-4 in terms of the fraction of the mass of all nuclei contributed by each type of nucleus. The graph shows the dependence of the abundance of each nuclear species on the present baryon density, the value of which specifies a particular big bang model. Figure 8-4 includes the range of values of the present baryon density that are compatible with observational evidence. The figure also shows the observed abundances of the elements, as determined by the number of atoms formed from these nuclei. These abundances represent averages obtained from various locations throughout our local region of the universe. Typical sources of these abundance determinations are meteorites (which bring us samples of the material that formed the solar system), spectra of the sun and other stars in our galaxy, and spectra of the light from the interstellar gas in our galaxy and in other galaxies.

Let us compare these observed abundances with those produced in our big bang model of the early universe, proceeding from the heaviest to the lightest nuclei. Theoretical models of stars as well as considerable observational evidence lead us to conclude that most of the heavy nuclei were created in the hot cores of stars. This conclusion appears consistent with the far smaller production of these elements in the early universe (see Figure 8-4, where the abundance marked "$A \geq 12$" refers to the sum of all nuclei containing 12 or more baryons). Of the group of somewhat lighter nuclei (^{6}Li, ^{7}Li, ^{9}Be, ^{10}B, ^{11}B), all except ^{7}Li appear to have been produced in collisions of cosmic-ray nuclei with the nuclei of interstellar gas atoms

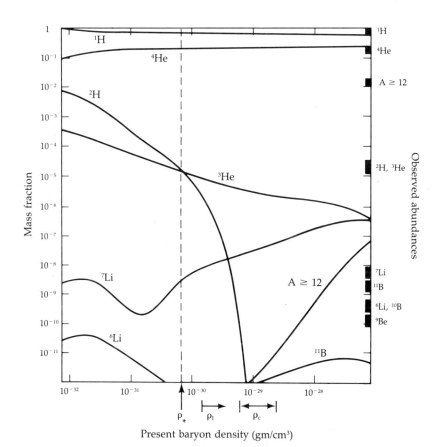

Figure 8-4. The final abundances of nuclei produced in the early universe (in terms of their fraction of the mass of all nuclei) are plotted against the only adjustable parameter that characterizes the various standard big bang models, the present baryon density (the mass density of neutrons and protons). The observed abundances of the light nuclei, as well as the observed total abundance of all those nuclei containing at least 12 nucleons (A ≥ 12) are indicated on the right. The density (ρ_*) of the particular model that produces the observed amounts of ^{2}H, ^{3}He, ^{4}He, and ^{7}Li is indicated, as are the ranges of possible values of the density of invisible matter (ρ_I) and the critical density (ρ_c), which separates closed and open universes.

during the present epoch. The abundances calculated from this well-understood process agree with the observed abundances of these four nuclei (^{6}Li, ^{9}Be, ^{10}B, ^{11}B). Furthermore, stars tend to destroy rather than produce these nuclei, and the calculated primordial production of these four nuclei is well below their observed abundances. (See Figure 8-4; the calculated primordial production of ^{9}Be and ^{10}B is less than 10^{-12}, and therefore does not appear on the graph.)

The remaining nuclei to consider are ^{7}Li, ^{4}He, ^{3}He, and ^{2}H (another symbol for deuterium). The abundance of protons, ^{1}H, is determined by the fact that the sum of the abundances equals one. A striking fact emerges from a close inspection of Figure 8-4. A big bang model universe whose present baryon mass density equals approximately 7×10^{-31} grams per cubic centimeter (indicated by the dashed line) would produce amounts of the above-mentioned remaining nuclei, and only these nuclei, in the observed abundance ranges. We shall denote this "favored" density by ρ_*. However, since stars may also be capable of producing the observed abundances of ^{3}He and ^{7}Li, we shall focus our attention on deuterium (^{2}H) and helium (which is primarily ^{4}He rather than ^{3}He).

The Importance of the Helium Abundance

We know that stars make helium as part of the process by which they produce energy. Observations suggest, however, that the bulk of the helium in the universe existed even before the formation of the stars that we see today. This observed "primordial" abundance (20 to 25 percent of the mass in all nuclei) can be produced by big bang models with a present baryon mass density in the range from about $\rho_*/20$ to ρ_*.

Such a result has great importance because it provides evidence that the big bang model remains valid at least as far back as the era during which helium nuclei were formed. Specifically, if we consider any other model of the early universe (which means violating at least one of the standard assumptions underlying the big bang model; see Chapter 7), we find that the theoretical abundance of helium usually attains a value outside the limits we have just imposed from observation. In addition, if the universe's lepton number were large, so that more neutrinos (or antineutrinos) than photons existed in an average region of space, the neutrinos would be degenerate (see Chapter 5) and the helium abundance would again change drastically. Finally, if new species of light leptons or bosons existed during the nucleon-to-nuclei transition era, more helium would have been made. For example, if new types of neutrinos are discovered with future accelerators, the theoretical helium production could exceed the observed upper limit on the helium abundance, in which case we would have to abandon the simple big bang model. In most such "nonstandard" models the helium production changes, because either the expansion rate of the universe or the initial value of the neutron-proton ratio differs from the value in the standard model.

The Importance of the Deuterium Abundance

Finally let us discuss the implications of the abundance of the lightest complex nucleus, deuterium. Stars easily destroy any deuterium present (by reaction 4a in Figure 8-3, which proceeds rapidly at stellar temperatures). The subsequent incorporation and expulsion of matter by generations of stars would therefore have destroyed some of the primordial deuterium. Hence the primordial abundance of deuterium could have been somewhat greater than its present observed abundance (shown in Figure 8-4). The same could hold true for ^{3}He. We therefore conclude from Figure 8-4 that the baryon mass density could be somewhat lower than the value ρ_*. But even though the value ρ_* represents only an upper limit to the present baryon density allowed by our models of the universe, this upper limit still has profound implications, if correct—that is, if the standard big bang model is valid, as the helium abundance implies.

What Can We Conclude?

The implications of this picture of primordial nucleosynthesis arise from a comparison of ρ_* with two other mass densities, that of the invisible matter (which we call ρ_I) and that required to "close" the universe (which we call the critical density, ρ_c). Note that ρ_* could represent either visible matter (in the form of ordinary stars) or invisible matter (if in the form of low-mass stars or black holes), since most of the mass of all such objects comes from baryons. Recall that the value of the critical density is determined by the value of the Hubble constant (H_o). Observations show that the "Hubble time" $1/H_o$ probably equals 10 to 20 billion years (see Chapter 7). If this is correct, the critical density (ρ_c) lies somewhere in the range shown in Figure 8-4, which is more than ten times the value of the baryon density implied by the light element abundances (at most ρ_*). Therefore, if baryons are now the major contributors to the mass in the universe, the universe must be open (with negative curvature) and will expand forever.

The second implication of primordial nucleosynthesis arises from the fact that the amount of invisible matter far exceeds the amount of visible matter. Although the density of visible matter appears to be less than the favored density (ρ_*), the density of invisible matter (ρ_I) seems to be at least twice as great as ρ_* (see Figure 8-4). If this indeed is true, then we reach a startling conclusion: The invisible mass (and therefore most of the mass of our universe) cannot have the form of ordinary matter (baryons). It must consist of some other type of particles, such as the massive neutrinos or magnetic monopoles we have already discussed. Note that if the second

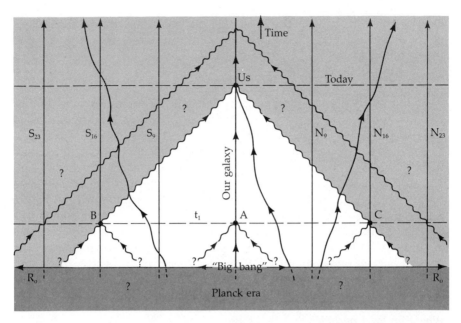

Figure 8-5. This space-time diagram depicts the histories (vertical "world lines") of particles at various distances from us in two opposite directions on the sky (corresponding to the right or left of the central world line depicting the history of our galaxy). The horizontal axis indicates the *present* distance (R_o) from us of those particles that follow the average motion of matter in the universe. For instance, the subscript on particles that we see in some northerly (N) direction and some opposite southerly (S) direction indicates their present distance from us in billions of light-years. Time proceeds (nonuniformly) in the vertical direction. The wavy lines indicate the histories of massless particles or signals that travel at the speed of light. The curved lines represent those particles or signals that move at a speed less than that of light with respect to the average motion of matter (vertical world lines). The region of the universe that we can know about (the clear triangular region) is bounded absolutely on the left and right by our "past light cone," the world lines of those signals traveling at the speed of light that reach us today. This region will be bounded at the bottom by the "graviton barrier" (labeled as the time of the big bang) until we know the laws of physics that existed during the Planck era. Note that the big bang occurred *everywhere* at a given instant of time.

implication (that the baryons do not dominate the mass of the universe) is valid, the universe may not be open (which followed from baryon dominance).

The Ultimate Horizon

Having now completed our survey of the history of matter, we again confront the basic question that hovers in the background: How much can we know about the universe? We can understand the ultimate limits to our knowledge by examining Figure 8-5.

Figure 8-5 (like Figure 7-4) is a space-time diagram. The vertical axis of this diagram represents time, which increases (although not uniformly) as we move upward. The central vertical line depicts our "world line": the history of the matter in our neighborhood. The dot labeled "us" represents the present moment of this history.

As we move away from our world line in either direction, we find the vertical world lines of other bits of increasingly distant matter. In Figure 7-4 the horizontal axis indicates the actual distance (R) of an object at any time in any particular direction on the sky. In Figure 8-5, the right-hand portion of the horizontal axis indicates the present distance (R_o) of an object in any particular direction on the sky. The left-hand portion of the horizontal axis indicates the present distance of an object in the opposite direction on the sky. Hence the horizontal axis provides a labeling of particles. For instance, the other straight vertical world lines shown in the diagram represent the histories of six bits of matter, at present distances of 9, 16, and 23 billion light-years in both the northerly (N) and southerly (S) directions. We use the labels N and S to remind ourselves that we are looking in opposite directions. Although the actual distances (R) of these bits of matter have been steadily increasing, their labels (R_o) have remained constant, as shown in Figure 8-5, since the labels denote their present distances from us.

The wavy lines in this figure depict the paths of signals that travel at the speed of light. They could therefore represent electromagnetic waves, gravitational waves, or massless neutrinos, for instance. The two wavy lines that meet "us" show the past history of the signals that we receive today from two opposite directions in space. Similarly, the wavy lines that intersect any bit of matter in the past (such as at the time t_1, at points A, B, and C) show the past history of the signals received by those bits of matter at that time. Our measure of the time coordinate has been chosen so that all the world lines representing travel at the speed of light make an angle of 45 degrees with the vertical direction.

Einstein's theory of relativity rests on the postulate that no signal can propagate with a speed greater than that of light. Hence the world lines of all signals (and thus of all particles) must make an angle of 45 degrees or less with the vertical direction. A few such histories of massive particles, sound waves, or anything else that travels at a speed less than that of light appear as the curved world lines in Figure 8-5. Their speed is measured with respect to observers who move with the average motion of the universal matter (represented by the vertical world lines).

Figure 8-5 also shows clearly why there is no "center" of the universe. As we move to the right or left in the diagram, no properties of the universe change. In particular, note that the big bang occurred *everywhere* at a given instant of time, not within an isolated region in space.

The Limits of Our Universe

What can we learn from Figure 8-5? Our most secure conclusion, and a fundamentally important one, is this: We can gain no knowledge whatsoever of a vast region of the universe. This inaccessible region (the portion of space-time shown by the upper-right and -left shaded areas) lies outside our "past light cone." Our past light cone, formed by the world lines of all those signals traveling freely at the speed of light that reach us today, bounds the space-time volume from which we can receive information. This four-dimensional volume is represented in Figure 8-5 by the (two-dimensional) clear area. (The other two dimensions correspond to different directions of viewing. If we restrict ourselves to a set of directions such as a circle around the sky, the resulting three-dimensional space-time volume resembles a cone.)

Signals, such as photons, that reach us along our past light cone can tell us only about a bit of matter at a particular time in its history, the time when the signal left that bit of matter (such as at event B or C). Hence as we look deeper into space we do see more and more matter, but we see that matter only at a particular time, earlier in its history. We can also receive information from any event (a point in space-time) within our past light cone. For instance, relic particles tell us about the past history of the matter in the vicinity of our own world line. But observations carried out at the present epoch cannot reveal any events that lie outside our past light cone. In particular, we cannot observe any other galaxies as they are today, since such space-time points are located at the intersections of the horizontal line "today" and the vertical world lines of the galaxies, outside our past light cone. Fantastically strange things could be happening in the universe today, but we have no way of finding out—until later.

The only way to look "beyond" this ultimate horizon is to let time pass. Then as the dot representing our present moves upward along our world line, our past light cone expands (as shown in Figure 8-5), allowing us to receive information from each bit of matter at a later time in its history. Conversely, we see that in our past (for example, at point A), a smaller space-time region of the universe was accessible to us.

For us (or for any other observer), the scientifically significant universe must be defined as all the events that took place within our past light cone. All other events, being undetectable, lie outside the scope of science. They can have no effect on our present or our past. They will affect our future, but at present, we cannot know how.

The existence of this horizon then leads us to ask: Does a limit exist to the amount of matter in this accessible universe? That is, does the region of space-time within our past light cone, which we may call our "causal universe," contain some portion of the world lines of *all* matter, or of only a portion of the matter in the universe? In the standard big bang model, as we extend our past light cone further and further back in time, it encompasses more and more matter (see Figure 8-5). At the Planck time, which we have somewhat arbitrarily chosen to call the time of the big bang, we see that the "view" along our past light cone has reached only a certain amount of matter. Moreover, this amount of matter is not much greater than the amount we can detect with our present telescopes.

However, if the unknown physical conditions in the early universe were sufficiently different from those we have assumed, the light cones might not remain straight as we follow them back in time, but could diverge to encompass much larger values of R_o as we approach the Planck era. Our past light cone would then contain more matter. Of course, we could also "view" more matter if signals could propagate to us through the Planck era, beyond the big bang. But since we understand virtually nothing about the nature of reality under such extreme conditions, it would be presumptuous to speculate in any detail about the universe during the Planck era. The Planck time represents another barrier to our understanding, but unlike the photon barrier, this is a theoretical barrier that we may hope to penetrate one day.

It is also possible that the universe did not emerge from such an extreme state as we have described. In principle, the collapse of the universe during a past phase could have slowed with time, eventually reversing to become an expansion. Such a scenario, in which the universe would "bounce" and thus never achieve the extreme densities and temperatures characteristic of the Planck era, involves universal acceleration rather than the deceleration that occurs now. The "bounce" could result from an effective density ρ that behaved so strangely during this remote era that the upper two curves in Figure 7-3 intersected at an extremely small value of the scale factor as well as at the point shown ($H^2 = 0$). But we know of no natural way to produce such a situation. Therefore, for now we must regard the Planck barrier as another effective limit to our universe.

Hence, within the standard big bang model, a limit does exist to the amount of matter in our causally connected universe. This limit is the amount of matter contained within our past light cone at the Planck time. For example, the bits of matter labeled S_{23} and N_{23} in Figure 8-5 lie outside our causal universe, since we have no way of knowing of their existence and they have no way of affecting us. Furthermore, as noted above, our causal universe used to be smaller, for example at event A in Figure 8-5. Note that signals from the matter labeled S_{16}, S_9, N_9, and N_{16} had then not had time to reach us.

The key point is this: The effective size of our universe (the amount of detectable matter) is determined by the fact that any signals have only had a certain finite time to propagate to us since the big bang. To observe, or to be influenced by, any matter in the universe, the signals from that matter must have had time to reach us. But since the speed of light absolutely limits the speed of any such signals, the product of the speed of light and the time since the big bang limits the present distance (R_o) of the matter from which such signals can arrive. For us, or for any other observer on any other world line, the effective size of the universe grows with time, as more and more matter comes into "view." As the universe emerged from the Planck era, the causal universe of any particular particle contained only a few other particles. By contrast, our causal universe now contains roughly 10^{88} particles.

Paradoxes of Causality

This limit to the size of the causal universe leads to consequences that disturb us because they make it difficult to understand the large-scale structure of the universe. The first problem concerns the isotropy of the microwave background radiation. In Figure 8-5, let t_1 be the time corresponding to the photon barrier, the epoch from which this radiation has propagated freely to us. (This time should actually appear much closer to the Planck time, but this does not affect our argument.) When we view the background radiation in two opposite directions on the sky, we observe the universe as it was at two widely separated points in space (points B and C). But note that the causal universes of the bits of matter (S_{16} and N_{16}) that emitted this radiation were completely distinct. No matter was contained in common within the past light cones of these two events (B and C). We may then ask, how did these two causally unconnected bits of matter manage to acquire the same temperature, which our observations of the microwave background reveal they possessed?

Our simple big bang model forces us to the conclusion that we can view regions of the universe that have not yet communicated with each other. Yet these regions are observed to have almost identical properties. Why should this be so? This is one of the paradoxes of causality.

Another paradox concerns the apparently antithetical problem of non-uniformity: the origin of large-scale structure in the universe. Let us consider small fluctuations in the density of the early universe, whose amplification by gravitational forces we believe eventually produced galaxies and other astronomical objects. At early times, when these fluctuations somehow developed, each causal universe contained far fewer baryons than needed to make a galaxy. The matter in our galaxy today was then contained within many causally unconnected universes. The question then arises: How did the fluctuation that developed into our galaxy (or any other) establish itself over regions that had not communicated with each other?

At present, the answers to these questions lie hidden in the unknown physics of the early universe. It is tempting to speculate that the simplest way to resolve these paradoxes of causality might also be the way that nature has chosen. That is, these paradoxes may show that our past light cone did extend further back in time and/or farther out in space than indicated in Figure 8-5 by the simplest big bang model.

"This is the part I always hate."

9

Where Do We Stand?

LET US STEP BACK to assess our current understanding of cosmic evolution and the many uncertainties that remain. The understanding we seek is contained within the answers to two encompassing questions: What are the properties of the universe? Why does our universe have these particular properties? We shall discuss both questions and assess what progress toward further answers we may anticipate.

Cosmic Evolution: A Personal View

In our exploration of the past history of the universe, we have made the cosmic clock run backward through time, with the equations of the big bang model to guide us on our journey. Let us now review what this theory tells us by letting time proceed forward. The picture that emerges is one of evolution on all scales, from the decelerating expansion of the universe itself to the microscopic dance of elementary particles as they annihilated and were created anew, eventually forming higher levels of structure. Evolution at intermediate scales of size appears to have been initiated by the gravitationally induced growth of tiny fluctuations within the general uniformity of the early universe.

To obtain a feeling for the nature of this evolution on a more personal level, let us trace *your* history—the evolution of that bit of matter that has now become "you." Basically, you are a complex arrangement of two kinds of elementary particles: quarks (of two types, up and down) and electrons. Also present within you—in the form of the forces that bind these particles together—are gluons and photons.

Primordial History

If the universe emerged from a symmetric state in which equal numbers of particles and antiparticles existed, then you owe your existence to whatever process broke that symmetry. At the very least, large domains containing slight differences between the number of baryons and antibaryons must have been created. The slight excess of baryons within the domain of interest to us (which is probably at least as large as the visible universe) can be expressed by the net baryon number, which remained constant after the asymmetry was established. Your body contains about 10^{29} baryons in the form of atomic nuclei.

The actual particles that make up your body did not exist until they were no longer subject to annihilation, which occurred after a mere blink of the eye—about a thousandth of a second—in the case of the quarks, which had by then formed neutrons and protons. A few hundred seconds later, as most of these neutrons were incorporated into a few light nuclei, the electron-positron annihilations ceased. A bit less than a million years after that, the protons and other nuclei combined with the surviving electrons to form atoms.

When microscopic evolution had proceeded to this point, macroscopic evolution began in earnest. Sufficiently large regions with higher-than-average density then collapsed under the pull of their own gravitation. This process triggered the formation of objects ranging in size from stars to the largest structures we observe, clusters of galaxies.

The object of interest to us in our present discussion, you, contains primarily oxygen (65 percent by mass), carbon (18 percent), and hydrogen (10 percent), with smaller amounts of nitrogen (3 percent), calcium (1.5 percent), phosphorus (1 percent), and other elements (1.5 percent). We have seen that the simplest element, hydrogen, has a primordial origin, but heavier elements such as carbon and oxygen were synthesized from the primordial nuclei (mostly hydrogen and helium) within stars. Nuclear fusion reactions produce heavier and heavier nuclei as stars evolve through states of higher and higher temperatures and densities. Furthermore, some stars can expel much of their mass into interstellar space after this nuclear processing occurs. In this way, the interstellar gas that pervades our galaxy has been continually enriched in these heavy elements.

Solar System History

Our story now shifts to the solar system, which formed from the collapse of a portion of this interstellar medium almost 5 billion years ago. As this cloud collapsed, it spun faster and faster, for the same reason that an ice skater spins faster when she pulls in her arms. The large rotational velocities of the outer part of this cloud prevented its further contraction toward the axis of rotation, but the collapse continued in the direction parallel to this axis and in the center of the cloud. The result was the formation of a hot ball of gas, the sun, surrounded by a rotating disk of gas, somewhat like a small-scale version of a spiral galaxy.

As the contraction of the sun proceeded to the point at which nuclear fusion of the sun's hydrogen began, the gas and dust in the surrounding disk began to condense into small, discrete objects. These objects increased in size through gravitational attraction of matter and through collisions with their neighbors. From these accretion processes came the planets, their satellites, and the smaller debris (asteroids, meteoroids, and comets) that now fill the solar system.

As the sun grew hotter, the temperature of the inner planets (Mercury, Venus, Earth, and Mars) rose to the point at which their gravitational forces could no longer prevent the escape of the lightest atoms, including the most abundant ones, hydrogen and helium. The hotter the temperature and the lower the mass of a particle in a gas, the greater will be its velocity. The lightest atoms and molecules had sufficient velocities to escape eventually from the gravitational fields of the hottest planets. Because of this escape, the four inner planets are mostly solid today, whereas the outer giant planets (Jupiter, Saturn, Uranus, and Neptune) are mostly gaseous and are composed primarily of hydrogen and helium, as are the stars. (Hydrogen and helium do not easily form solids or liquids.) The most distant planet, Pluto, more like a satellite than a planet, is an exception. Fortunately for us, the earth retained some hydrogen atoms, locked in relatively heavy molecules such as water.

Because of the large amount of hydrogen initially present, our primitive atmosphere probably consisted of hydrogen atoms bonded in molecules to the next most abundant atoms. This bonding produced hydrogen molecules (H_2), water (H_2O), ammonia (NH_3), and methane (CH_4). (Helium atoms do not form molecules.) Experiments have shown that when energy is supplied to a gas composed of these simple molecules, something remarkable can occur. Within the gas, relatively complex molecules, and in particular some of the amino acids and nucleotides that form the basis for all life on earth, appear. This process occurs regardless of the source of energy—electrical discharge (lightning), ultraviolet photons (primeval sunlight), or motion (such as the pounding of surf on a shore). Scientists believe that volcanic activity produced sufficient water to create the necessary aqueous environment and atmospheric shielding from the solar ultraviolet radiation that otherwise would have destroyed these complex molecules. (The atmospheric shielding results primarily from ozone [O_3], produced by the ultraviolet breakup of water molecules.) The combination of water, these other molecules, and any one of a number of possible energy sources should therefore have been sufficient to allow life to arise spontaneously on earth.

Scientists estimate that life emerged in this manner from the primordial soup some 4 billion years ago. Fossil records show that about 2 billion years later, microscopic blue-green algae were actively converting another product of volcanoes, carbon dioxide (CO_2), into molecular oxygen (O_2) by the process of photosynthesis. The presence of oxygen in the atmosphere then allowed oxygen-users, animals, to evolve through the process of natural selection.

Although various steps in this process could prove to be different from those just outlined, our chief conclusion from this brief survey of the evolution of matter on earth will remain: All forms of life we know of require heavy elements such as carbon and oxygen. You consist primarily of heavy elements that were synthesized inside stars that exploded more than 4.6 billion years ago. In addition, most of the hydrogen nuclei in your body are pristine relics of the still more distant past. The stars themselves are also relics of this remote time, since the seeds of their evolution presumably existed in the early universe. Hence if we are "children of the stars," we may also be "grandchildren of the primeval plasma."

Cosmic Evolution: The Major Questions

In any branch of science, as we probe more deeply, new questions arise from the knowledge just acquired. As we learn more, we are able to ask more penetrating questions. Each new frontier of science has its own characteristic set of problems. Given our present knowledge of cosmic evolution, let us now review some of the major questions that confront us today.

The Validity of Our Model Universe

Is the big bang model valid? In a sense, this is the most pressing question, because the big bang model provides the framework that guides most of the present research in cosmology. Recall that four assumptions form the foundation of the simplest, or standard, model (see Chapter 7). We know that the first assumption, the validity of general relativity, must break down at extremely high densities (during the Planck era). But since we do not yet know what theory replaces general relativity under such conditions, we do not know what effects any such breakdown would have on the subsequent evolution of the universe. At present, the best we can hope for is that the big bang model is valid after the Planck era.

We know that our second assumption, the Cosmological Principle, applies only to properties of the present universe averaged over distances greater than a few percent of its visible size. Was the early universe also uniform and isotropic on much smaller scales of size and mass, as we have assumed? We do not know, although we can set some limits on the degree of nonuniformity or anisotropy.

The validity of the third assumption, that antimatter is nearly absent in the universe today, can be strengthened by further searches for the various forms of evidence of the presence of antimatter. The final assumption, that most of the mass of the universe consists of the known types of elementary particles, may have been invalid only in the very early universe, at energies inaccessible to present experiments.

We have not considered in detail any alternate cosmological models, except for the cold big bang variation discussed at the end of Chapter 7. One of the best-known alternates is the *steady state model*, first proposed by Hermann Bondi, Fred Hoyle, and Thomas Gold in 1948. In this model, the properties of the universe remain constant in time because the continuous creation of matter replenishes the density of the universe, which would otherwise be reduced by its expansion. However, it is difficult to reconcile the lack of evolution of the overall properties of such a universe with the properties of quasars and of other remote objects that appear different when we observe them as they were further back in time. The deathblow to the steady state model seems to be the establishment of the blackbody spectrum of the isotropic microwave background radiation, since the model has no high-density phase to produce such a spectrum.

Another alternate model, considered at various times since it was first proposed by the Swedish astronomer C.V.I. Charlier in 1908, is the *hierarchical model*. In such a universe, the clustering of matter that we observe up to the scale of thousands of galaxies is assumed to continue to all scales of distance. However, the degree of clustering decreases as the amount of matter included increases. The detailed properties (in particular the evolution) of such a model have never been completely worked out. Moreover, the model appears to conflict with the observed degree of isotropy of the microwave background radiation, the strongest evidence for the validity of the Cosmological Principle.

Various other models of the universe have been proposed and are still being considered. However, very few of these models can produce in a natural way the observed abundance of deuterium nuclei. (Deuterium may furnish the key abundance because it is the only nucleus that cannot be made in stars or in any other known objects; hence an early universe with certain particular properties is required to produce it.)

The Values of the Universal Parameters

What are the properties of the present universe? The cosmologically significant characteristics of the present universe fall into three classes. The first class includes the universe's *dynamical properties*: its expansion rate (H_o) and its deceleration, expressed by the deceleration parameter (q_o). Recall that we can obtain the values for these quantities by determining the distances to remote objects whose red shifts have been measured. However, we still have no reliable and accurate distance indicator, the lack of which produces uncertainty within a factor of two in the value of H_o and a much greater uncertainty in the value of q_o.

The second class of characteristics includes the universe's *material properties*: the density of all forms of mass-energy (ρ_o) and the corresponding pressure (p_o). Recall that we can completely determine only that portion of the present density that is gravitationally bound into separate systems with visible matter (and hence is nonrelativistic). The pressure of such nonrelativistic forms of mass-energy is cosmologically insignificant. Any uniform component of the present density (such as a gas of relativistic particles) reveals itself mainly through its contribution to the deceleration parameter, unless of course we can observe the component directly. A gas of relativistic particles has a pressure equal to $1/3$ of its energy density.

The third class of properties consists of our determinations of the *ages* of various samples of the universe, such as stars and radioactive material. The ages of these objects provide us with a lower limit to the age of the universe.

Obtaining the values of these five properties—H_o, q_o, ρ_o, p_o, and the minimum age of the universe—is of critical importance for two reasons. First, these numbers furnish a test of the standard big bang model, which predicts certain relationships among them. For instance, we have seen that the present universe is predicted to be dominated by nonrelativistic matter (see Figure 8-2), in which case its pressure must be negligible. In addition, the gravitational field equations relate the material properties of the universe to its dynamical properties, and its dynamical properties to its age. Second, from these numbers we can determine both the spatial curvature and the future evolution of the universe.

The Nature of the Invisible Matter

We do know that invisible matter dominates the universe. The bulk of the mass of galaxies, of clusters of galaxies, and of the entire universe reveals itself not through starlight but only through the gravitational field it produces. What comprises the invisible matter? Methods for determining the nature of this matter, whether it be in the form of massive neutrinos, black holes, faint stars, or other "exotica," are now being actively explored. Of course, uniformly distributed invisible matter could also exist. For instance, if the lepton number of the universe were sufficiently large, the universe could be dominated by degenerate massless neutrinos. However, this would drastically alter the abundances of the nuclei produced in the early universe.

The Origin of Structure

Our universe has a marvelous complexity, revealed on scales that range from the diverse forms of life to the largest astronomical objects. Yet all indications point to a universe that was much simpler in the distant past. For example, the cosmic microwave background is, to a high degree, featureless. This is consistent with our belief that fluctuations in the universe tend to grow with time, resulting in a universe that was smoother in the past. Once large-scale structures such as galaxies had developed, a cascade of smaller-scale structures, such as stars, planets, and at least in one case, life, developed within them.

The basic question then becomes apparent: How was this structure, in the form of small fluctuations, imprinted on the early universe? This question—the origin of structure—leads in turn to another: Why was the early universe so nearly uniform? We have seen that both of these problems become much more difficult to resolve if each bit of matter was influenced by only a small amount of surrounding matter at early times, as is the case in the standard big bang model.

Could the very early universe have been extremely chaotic (rather than smooth), as postulated by some scientists who regard this as a more natural "initial condition"? Investigations of models of this type show that such universes would have great difficulty "smoothing themselves out" in time to make the observed amounts of deuterium and helium, or to resemble the present universe. Thus we face what we may call "the enigma of evolution": The seeds of all the structure we observe must somehow have been planted as small fluctuations in the properties of the primeval plasma. We remain ignorant, however, of the process that created these fluctuations.

A related question concerns another fundamental property of our universe: the relative amounts of its basic constituents—the types of elementary particles that are, and were, present. For instance, what was the origin of the small ratio of baryons to photons that we observe today?

In the section that follows we shall consider some of the ways in which we hope to proceed toward answering these difficult questions that confront us.

Prospects for Future Understanding

We have discussed certain difficulties inherent in cosmology and astronomy, which arise because we cannot perform experiments on the objects of interest; we cannot bring them into our laboratory for study. These difficulties become most severe at the frontiers of cosmology, where even observations become, in some cases, impossible. A greater burden then falls on the third component of the scientific method, theory. For example, the origin of structure and the laws of physics during the Planck era are two problems that at present can be attacked only through theoretical analysis. We must employ theoretical models in lieu of experiments, letting the computer serve as our laboratory, in which we investigate how distant objects, or the universe itself, behave under various conditions. The traditional roles of theory, to explain our observations and to suggest new ones, become especially critical in cosmology because of the relative paucity of observational data.

Nevertheless, true progress in our understanding of the universe will be governed by our ability to increase our observational capabilities. In the realm of direct data, we must extend the view of the cosmos that we obtain via photons and broaden this view to include other direct carriers of information such as gravitons and neutrinos. In the realm of indirect data, we must continue our search for the ashes of our past, the various types of relic particles from the early universe.

Detection of Cosmic Photons

Since most of our information has in the past arrived, and for the forseeable future will continue to arrive, in the form of photons, let us briefly assess the progress we may expect in this arena. First, we must understand in more detail the nature of such observations.

A beam of radiation from a distant source is characterized by its wavelength (or frequency, or photon energy), its polarization (which need not concern us here), its location on the sky (angular position), and its flux (amount of energy per unit area of the beam that strikes the detector per unit time) per unit frequency. Recall that the flux (brightness) received from a particular object is proportional to its intrinsic luminosity divided by the square of its distance. Actually, the distance to an object is not uniquely defined in cosmology; it depends upon what method we use to measure it. For instance, if we could use yardsticks laid end to end from us to an object at some fixed cosmic time, should its distance be that measured when it emitted the light or that measured today, when we receive the light? Because of this ambiguity, astronomers prefer to use the corresponding directly observable quantity, the object's red shift, which is proportional to all measures of distance for nearby objects. In any case, in order to look farther out in space and further back in time, we must either discover more luminous sources or develop methods to detect fainter ones.

The discovery of quasars represents a perfect example of the first way we can extend our vision. Quasars could have been discovered years earlier, if astronomers had realized that the light that forms their starlike images at visible wavelengths is in fact highly red shifted. However, not until intense radio waves were detected from their precise locations on the sky was the peculiar nature of quasars discovered. Visible-light spectra of these apparent "stars" then revealed their large red shifts. (Before the quasars were found to be "peculiar," astronomers did not have any particular reason to spend hours obtaining their spectra.) In order to produce the light we detect at such large distances from the quasars, they must have enormous intrinsic luminosities.

We now turn to the second way we can extend our vision, the one upon which we must usually rely. We can improve our ability to detect fainter sources in three basic ways. First, we can detect smaller fluxes of photons by increasing the collecting area of our telescope, since it is the flux multiplied by the area over which we collect it that determines the number of photons received per second. Second, we can improve the efficiency of our detectors, so that a greater fraction of all photons received is actually detected. Then the same number of photons received per second can produce a detectable signal where none existed before.

Finally, we can reduce the noise that tends to obscure the signal we are trying to detect. Noise can be radiation from other sources that cover all or part of the sky (the "background"), or radiation from the telescope itself, or spurious signals within the detector. The major sources of background radiation are (1) distant sources in the same direction as the object we are studying, and (2) the scattering and emission of radiation by particles in our atmosphere, in the interplanetary medium, and in the interstellar medium. We can reduce the amount of noise in various ways. First, if we are trying to detect objects with extremely small angular sizes, rather than sources that cover an extended area of the sky, we can sharpen the focus of our telescope so that radiation from a smaller fraction of the sky enters our detector. Then the amount of background radiation will be similarly reduced. If the angular resolution of our telescope remains larger than the angular size of the source, the signal from the source will remain the same, since all of the signal will still enter the detector. In most cases, the angular resolution of a telescope improves in direct proportion to the telescope's size. Hence this improvement in focus represents another way in which an increase in the collection area helps us detect fainter sources.

A second way to reduce the noise is to cool the telescope so that it emits less radiation. Such cooling proves useful for infrared astronomy, since objects at typical terrestrial temperatures radiate most of their energy at infrared wavelengths. Cooling the detectors also increases their efficiency, since it reduces their internal thermal noise.

A third way to reduce the noise is to place the telescope in orbit above the earth's atmosphere. This eliminates the background radiation from our atmosphere. More importantly, it reduces the size of the images of "point" sources such as stars, since the atmosphere no longer scatters light from the source. This in turn allows the competition from other sources of background radiation to be reduced by sharpening the focus of the telescope, as mentioned above. An even more important benefit of observations from space arises from the fact that although information from the universe comes to us at all photon wavelengths, our atmosphere absorbs all but the narrow band of visible and some infrared wavelengths and the broader band of radio and some microwave wavelengths.

Figure 9-1 shows how far we have progressed in our exploration of the universe by observing radiation of all available wavelengths. The left-hand vertical axis of Figure 9-1 indicates red shift, a measure of distance; the right-hand vertical axis indicates the corresponding time that the photons were emitted, as measured from the big bang. The clear area indicates the range of epochs (or red shifts) from which photons of each wavelength can propagate to us. The dark regions within the clear area indicate the extent to which astronomers have already explored the universe. The left-hand boundary of the clear area exists because photons with sufficient energy can annihilate upon collision with the photons in the microwave background, producing electron-positron pairs ($\gamma + \gamma \rightarrow e^- + e^+$). Hence these high-energy photons cannot reach us from distant sources. The right-hand boundary of the clear area (the right-hand vertical axis) exists because radio photons with wavelengths greater than about a million (10^6) centimeters are absorbed by the free electrons in the interstellar medium of our galaxy. The upper boundary of the clear area in the diagram is the photon barrier. The lower boundary of this area (the lower horizontal axis) represents the distance (corresponding to a red shift of 0.01) at which the universe begins to appear uniform and the velocity of galaxies arises primarily from the universal expansion: the beginning of the realm of cosmology. Let us now use Figure 9-1 to discuss the limits to our current view of the universe in each wavelength range, and the prospects for extending that view in the future.

Gamma-Ray and X-Ray Radiation

At the shortest wavelengths, rocket and balloon flights provided us our first glimpse of what the nearby universe looks like in the "light" of gamma rays and X rays. Since that time, observations at these wavelengths made from earth-orbiting satellites have revealed a great variety of phenomena. Among the more unexpected results were the first evidence for the existence of black holes (see Chapter 5) and the discovery of objects (probably neutron stars) that emit short, intense bursts of gamma rays and X rays. Of cosmological importance is the background radiation that has been found to cover the sky at these wavelengths. This background radiation is *not* the same as the cosmic microwave background, which contributes no flux at these short wavelengths. The source of this background radiation remains unknown, but we do know that some fraction of it comes from violent objects such as quasars.

The Einstein X-ray Observatory, launched into earth orbit in 1978, has detected X rays from extremely distant quasars as well as from the hot gas within distant clusters of galaxies. The limit corresponding to the upper boundary of the dark region in the X-ray range in Figure 9-1 represents the distance to the farthest quasar detected at each X-ray wavelength.

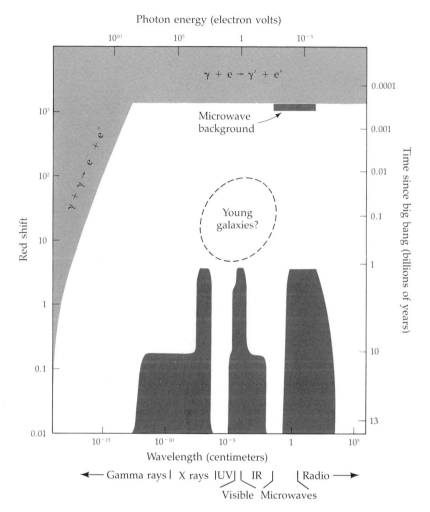

Figure 9-1. This figure shows the extent to which we can explore the universe throughout the spectrum of electromagnetic radiation (in terms of either the red shift of sources or, equivalently, how far back in time we see them). The darkly shaded areas show the extent of our present knowledge. The lightly shaded area shows the region that we can never view directly because the photons are either scattered by electrons ($\gamma + e \rightarrow \gamma' + e'$) or collide with other photons, producing electron-positron pairs ($\gamma + \gamma \rightarrow e^- + e^+$). The dashed boundary surrounds the region where we may see galaxies in their early phase of development.

However, it is likely that at least some of the background radiation comes from still greater distances. We have been unable to penetrate equally far into the universe at the shortest wavelengths shown in Figure 9-1 because the flux of photons from cosmic sources decreases with increasing energy. Most sources do not emit enough high-energy (gamma-ray) photons to be detectable with our present equipment.

Among the projects planned for the future at such wavelengths, the two that will probably prove of greatest value to cosmology are the Gamma-Ray Observatory (GRO) and the Advanced X-Ray Astronomy Facility (AXAF). These telescopes, like all the others that will operate from space, will be placed into an earth orbit, presumably by the Space Shuttle. Their increased sensitivity will allow us to extend our view of the universe in X and gamma rays beyond the limits shown in Figure 9-1.

Ultraviolet and Visible Radiation

As we move to longer wavelengths, we find a second significant gap in our ability to probe the universe: the ultraviolet range. Ultraviolet photons with energies ranging from 13.6 electron volts to a few hundred electron volts can ionize atoms in the interstellar gas and are absorbed in the process. As a result, few such photons from any sources outside our galaxy can ever reach us. At longer ultraviolet wavelengths, which correspond to photon energies from about 4 to 13.6 electron volts, our galaxy becomes transparent, but our atmosphere remains opaque. Observations of a few extragalactic objects have been made from orbiting telescopes with detectors sensitive to these longer ultraviolet wavelengths.

However, in 1985 the Space Shuttle is scheduled to place a major new telescope into orbit. This Space Telescope (Figure 9-2) will detect photons with energies ranging from about 1 to 10 electron volts. In addition to the longer ultraviolet wavelengths, these energies cover the visible range. Because the Space Telescope is larger than previous ultraviolet telescopes, it will be able to see farther at such wavelengths. But the Space Telescope has additional advantages in the visible range. The elimination of atmospheric scattering and emission means that point sources will appear smaller, and background noise will be reduced. As a result, we shall be able to observe objects 100 times fainter than those we can see from the ground. This represents a major advance in our ability to view objects at greater distances, and hence at earlier times. In addition, this telescope will sharpen our view and thereby help to increase our understanding of a great variety of less distant objects.

Nevertheless, ground-based telescopes will retain one important advantage over the Space Telescope at visible wavelengths. They can collect more photons, because the largest terrestrial telescopes already have mirrors much larger than the mirror of the Space Telescope. Furthermore, the con-

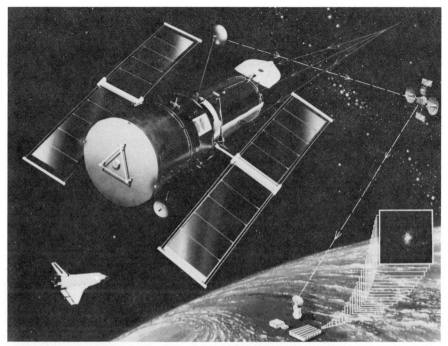

Figure 9-2. The Space Telescope is scheduled to be placed into earth orbit by the Space Shuttle (shown at the lower left) in 1985. Digitized data from the ultraviolet and visible radiation collected by the telescope's 2.4-meter-diameter mirror will be beamed to ground stations, where it will be converted to pictures. With this telescope, we shall be able to expand our view of the universe to include objects 100 times fainter than can be seen with ground-based telescopes. (Photo: NASA, Marshall Space Flight Center.)

struction of still larger mirrors (or groups of mirrors with a common focus) has begun. For certain types of investigations, ground-based telescopes will therefore remain superior, and hence will complement the abilities of the Space Telescope.

Figure 9-1 also shows the extent of our view at visible wavelengths, determined by the most distant individual object yet observed, a quasar with a red shift of 3.78. (From the general formulae given in Chapter 7, one plus the red shift equals the ratio of received to emitted wavelength; it thus also equals the universal scale factor today [one] divided by the scale factor when the radiation was emitted. All matter was therefore 4.78 times closer together when the light left this quasar.) Observations suggest that quasars do not exist at much higher red shifts, that is, at much earlier times. Nevertheless, we expect to see young, luminous, still-developing galaxies at large red shifts, as shown in Figure 9-1. These galaxies may completely cover the sky, since they were much closer together at the early epoch in cosmic history at which we shall see them. Their red shifts may be so large that their light reaches us not at ultraviolet or visible wavelengths, as it was emitted, but instead in the infrared (IR) region of the spectrum.

Infrared and Microwave Radiation

Except for some narrow bands at the shorter wavelengths, our atmosphere absorbs infrared photons. In addition, our surroundings emit radiation copiously at infrared wavelengths. To reduce these problems, clever techniques for observing celestial objects, such as from a high-flying airplane with a supercooled telescope, have been developed. Still more significant gains in observational ability will follow the launch of the Infrared Astronomy Survey (IRAS) satellite, which will house a telescope cooled to a few degrees above absolute zero. The proposed Shuttle Infrared Telescope Facility (SIRTF) would further increase the range in distance and wavelength we can observe, and it would allow astronomers to examine in detail the many objects that will be discovered by IRAS. If these missions succeed, the gap in the infrared region of Figure 9-1 should fill dramatically. Once again, these new telescopes will also help us to solve many "local" problems, such as how stars form from the interstellar gas and dust. In addition, they might be able to detect the universe's invisible mass if it has the form of low-mass stars, which emit most strongly at infrared wavelengths.

In the microwave region of the spectrum, we find quite a different situation. The detection of the microwave background radiation has already allowed us to "see" as far as we ever can, back to the photon barrier (see Figure 9-1). What we need now are more accurate measurements of the spectrum of this background radiation and of any deviations from its almost uniform distribution over the sky. These observations can tell us what structure the universe had acquired by the time that atoms formed, a few hundred thousand years after the big bang.

All the early measurements of this 3-degree blackbody background radiation were made from the earth's surface. Because our atmosphere emits microwave radiation, earth-based observations are possible only at wavelengths longer than a few millimeters. But most of the microwave background radiation exists at wavelengths shorter than a few millimeters (see Figures 4-4 and 6-6). Rocket and balloon flights have reduced the problem of atmospheric emission to some extent, but observations from space will be necessary to measure accurately the cosmic background radiation at all wavelengths.

The launching of the Cosmic Background Explorer (COBE) satellite, shown in Figure 9-3, will begin this important task. This satellite will be able to

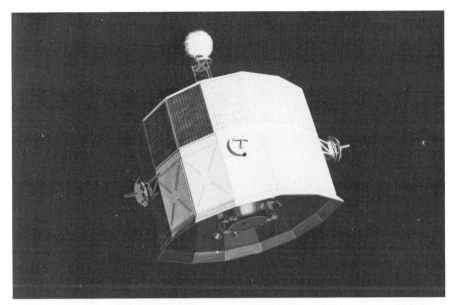

Figure 9-3. The Cosmic Background Explorer (COBE) satellite will contain liquid-helium-cooled detectors (shielded from stray radiation) that will measure the spectrum and distribution of the microwave background radiation and of cosmic radiation at shorter infrared wavelengths. (Photo: NASA, Goddard Space Flight Center.)

test accurately the big bang model's prediction that the background radiation has an almost perfect blackbody spectrum, and it should be able to detect the larger-scale deviations from isotropy of the radiation predicted by theories of galaxy formation. In addition, COBE will scan the sky at wavelengths shorter than those of the microwave background radiation, providing our first overall view of the universe at the longest infrared wavelengths. Because of its small size, however, COBE can collect photons only from fairly wide areas of the sky and therefore will be relatively insensitive to radiation from sources of small angular extent. To remedy this focusing deficiency, astronomers hope for the eventual launch of a large deployable reflector, capable of detecting point sources of microwave radiation.

Radio Radiation

At radio wavelengths, the detection of thousands of strong radio sources—quasars and radio galaxies—has allowed us to view the universe at distances comparable to those we can reach at visible wavelengths. Furthermore, extremely high angular resolution, greater than that available at any other wavelength, has become possible. We know that larger telescopes produce sharper images. An improvement in focus can also be achieved by interferometers—individual telescopes separated by large distances. Combining the signals detected by each telescope allows the array to act in a sense like a single telescope of a size equal to the extent of the array. By simultaneously observing a radio source with a network of telescopes widely dispersed around the world, radio astronomers have achieved incredible angular resolution. This resolution amounts to the ability to see the width of a hair on the head of a person in San Francisco from the top of Hoover Tower on the Stanford University campus, 45 kilometers away. With this remarkable ability, radio astronomers can observe regions almost as small as the solar system in nearby galaxies. The sources of the violent outbursts of particles that produce the great amounts of radio emission from some galaxies are confined to regions of this size.

In order to see to large distances, radio astronomers also require telescopes with large collecting areas. The most recent example of progress in this direction is the Very Large Array (VLA), shown in Figure 9-4. The many large reflectors, or "dishes," give the VLA the ability to collect large numbers of radio photons, and the large distances over which the dishes can be moved along their tracks allows them to produce an angular resolution comparable to that of the Space Telescope. The rotation of the earth also moves the dishes, allowing the VLA to achieve the focusing ability of a single radio telescope with a 27-kilometer diameter. The system has been operational for only a few years, but it has already provided the most detailed pictures of the multiple quasar images produced by gravitational lenses (see Chapter 6), as well as of a variety of other phenomena.

Finally, we note that the lack of observations at wavelengths longer than 2,000 centimeters (see Figure 9-1) arises from the fact that such photons cannot penetrate the earth's ionosphere. Therefore we shall require radio telescopes in space in order to view the universe at wavelengths from 2,000 to 1 million centimeters.

The Challenge of the Future

We can tell from Figure 9-1 that a vast portion of the universe observable via photons remains to be explored. We cannot easily predict what we shall find there, but we expect that we shall discover remarkable and unexpected phenomena, just as we have in the past whenever we have extended our vision, either in distance or in wavelength. We shall learn more about how the processes of nature operate under a great variety of conditions today, as well as how they operated in the distant past. The near future promises to be an exciting time of discovery, if we maintain the support needed to continue this exploration of the frontiers of our cosmic environment.

Figure 9-4. The Very Large Array (VLA), located near Socorro, New Mexico, consists of 27 large radio telescopes that move along railroad tracks in a Y-shaped pattern. The radiation collected by each "dish" is combined coherently with that collected by the other dishes and is computer analyzed to produce radio maps of distant objects with unparalleled clarity. (Photo: National Radio Astronomy Observatory VLA Program.)

Can astronomy benefit us in ways other than the intellectual or aesthetic? Physicist Charles Townes of the University of California at Berkeley has put forward an interesting argument that it may. In the early 1950s Townes developed the maser, for which he shared the 1964 Nobel prize in physics. A maser (microwave amplification by stimulated emission of radiation) contains atoms in some high-energy state. When they are irradiated with photons of the precise energy released in one of the atomic transitions to a lower-energy state, the atoms are stimulated to undergo such transitions, thereby adding photons of the same wavelength to the incident beam of stimulating photons. This process can produce an extremely narrow, intense beam of radiation. The same principle led to the development of the laser (light amplification by stimulated emission of radiation).

In the 1960s, astronomers discovered amazingly intense point sources of radiation at a few particular radio and microwave wavelengths that correspond to transitions between energy levels of certain molecules in the interstellar medium of our galaxy. They soon realized that the maser process is responsible for the emission from these cosmic sources, believed to be located in dense clouds surrounding certain types of stars. Townes notes that if microwave astronomy had developed some two decades earlier than it did, the process responsible for masers and lasers (and all their applications) would have been discovered in space, rather than in the laboratory. How many phenomena useful to us on earth are waiting to be discovered by astronomers? We do not know; only better instruments and serendipity will tell.

Why Is the Universe the Way It Is?

Behind the questions that deal with the properties of our universe lurks another, quite different type of question: Why does the universe have *these* particular properties? For other systems in nature, we deal with such a question in a standard scientific manner. The laws of physics govern how the system (a star, a planet, or a person) evolved from some specified initial condition. For a star, the initial conditions include the mass, composition, and rotation of whatever portion of an interstellar cloud eventually collapsed to form the star. For life on earth, the initial conditions include the chemical composition, temperature, and density of the primordial atmosphere and oceans, as well as the incident flux of solar radiation. Given the laws of physics, the question of why a system has certain properties thus becomes the question of why the system had the particular initial conditions that evolved to produce those properties.

This process cannot continue indefinitely. The buck stops when we come to cosmology, for we can no longer find a larger system to explain the "initial conditions" of the universe. In principle, we can understand the properties of all smaller structures within a particular model of the universe, since their initial conditions arise from evolutionary processes within that model. But what determined the initial conditions of our universe?

Even the meaning of the phrase "initial conditions of the universe" is unclear, especially since we cannot easily prescribe what we mean by the "beginning" of the universe. Rather than deal with such remote epochs, let us instead focus on certain fundamental properties that characterize the universe as it exists today.

The Mysterious Ratios

The first characteristic involves an apparent coincidence that has intrigued many cosmologists. The ratio of the size of the visible universe (which roughly equals the speed of light times the expansion age of the universe) to the size of an atom (which is determined by the mass and charge of the electron and by Planck's constant) is enormous, about 10^{36}. Another fundamental number arises from quantities that apparently have nothing to do with cosmology. This number is the ratio of the electromagnetic to the gravitational force between two protons (which depends only on the mass and charge of the proton and on the gravitational constant). A striking fact emerges: This second ratio equals 1.24×10^{36}. Why should the size (or age) of the present universe be such that these two huge numbers are comparable? Is this merely a coincidence? Or does the near-equality of the two ratios reflect some deeper connection between the laws of physics and the properties of our universe? Theories have been constructed to explain this equality, but none has proven completely viable.

The Curvature of Space

A second intriguing characteristic of our universe is its small spatial curvature compared to the huge range of possible values (either positive or negative). By referring to the universal evolution equation in Figure 7-3, we can better understand the significance of this property. Observations show that at present the expansion rate term (the left-hand side of this equation) is comparable to the density term (the first term on the right-hand side). The curvature term (the second term on the right-hand side) can therefore be, at most, comparable in magnitude to the density term today. Now recall that as the universe expands, the density term decreases more rapidly than does the curvature term. Hence in the distant past the density term must have been far larger than the curvature term and therefore must have almost exactly equaled the expansion rate term. From a Newtonian viewpoint, this is equivalent to the statement that the kinetic energy of expansion (proportional to the expansion rate term) is almost exactly equal to the energy of the gravitational attraction of the matter (proportional to the density term). Why was the universe so finely tuned in the past? Is it because its curvature is exactly zero? If so, why? Interesting proposals for dealing with these questions, involving the properties of the early universe, are being studied actively.

The Large-Scale Uniformity

A third important characteristic of our universe is that it obeys the Cosmological Principle over large distance scales. At least two facts make it difficult to understand why our universe is so uniform. First, only an infinitesimal fraction of all the models of the universe that we can construct within the framework of Einstein's theory of gravitation are as smooth as the actual universe. Second, as mentioned in Chapter 8, in the standard big bang model well-separated regions of our universe have not yet been able to influence each other because signals have not had time to travel between them. How could these separate regions have acquired the same properties? This problem provides another indication that the standard model may not have been valid in the distant past.

Another key characteristic of our universe is the ratio of photons to baryons. We shall not consider this ratio here, because it now appears likely that it can be expressed in terms of the fundamental constants of nature, whose values we are not seeking to explain. The relationship of this ratio to the more fundamental constants of nature can be derived if the photons have a stellar origin, with a different relationship appearing if instead the baryon number arises in the very early universe.

Can we explain the size, the near flatness (small curvature), and the uniformity of the visible universe? There have been some promising recent attempts within the framework of nonstandard big bang models. At present, our answer is still "no," since these particular properties of our universe have not yet been shown to be a unique consequence of a theory of possible universes. They simply correspond to a particular choice among models of the universe and to the particular value of its present age.

The Anthropic Principle

A quite different approach to understanding why the universe is the way it is deserves mention, since it attempts a comprehensive explanation (at least in the philosophical sense) of the universe's properties on the basis of a single proposition called the Anthropic Principle. According to this principle, the universe is the way it is because *only in such a universe could intelligent life have developed to observe the universe*. That is, the universe has its particular properties because if it did not, we would not be here to observe them. The Anthropic Principle is an arresting and provocative proposal. Let us examine it more closely.

The principle presupposes a definition of intelligent life and of the physical conditions required for its existence. This immediately introduces a large degree of uncertainty, because from our parochial viewpoint, any characterization of "intelligent life" is likely to be much too narrow. Nevertheless, with all appropriate disclaimers, certain general conditions appear necessary for the development of intelligence.

Intelligence involves the ability to store and to manipulate large amounts of information. This requires a complex system: many atoms forming a variety of structures, beginning with molecules. The existence of a variety of molecules in turn seems to require two other conditions. Since hydrogen alone cannot form many different molecules and since helium atoms do not bind to other atoms, some heavier elements are necessary. But at temperatures so high that the kinetic energy of the colliding molecules exceeds the energy required to disrupt them (their binding energy), molecules cannot exist. A sufficiently intense flux of other high-energy particles (such as photons) would also prevent the formation of complex structures. Yet if thermal kinetic energies were much less than the binding energies of molecules, the reactions between atoms and molecules would occur too slowly to build up complex systems in the available time, the age of the universe. We should therefore expect, as is the case, that the average temperature on earth corresponds to thermal energies of the same order as the binding energies of molecules. Complex molecules form, interact, and change their identity in great profusion on our planet.

These conditions—the presence of elements heavier than helium and a restricted range of temperature—do not *require* the existence of planets or even of stars. What do they require of the universe (in terms of the three basic properties on which we have focused)? To answer this question, we must consider an infinite number of possible universes, and determine which ones allow the development of complex structures. Since we cannot yet do this with any degree of certainty, we shall only sketch some of the arguments that have been advanced. Models that deviate strongly from homogeneity have not been studied in any detail, so we can draw no conclusions regarding the growth of complex structures in such universes.

Some degree of nonuniformity does seem to be required in order for complexity to develop. But in a universe expanding so rapidly that the kinetic energy of expansion far exceeds the energy of the gravitational attraction of its constituents, bound systems such as galaxies could never form. On the one hand, this places a strong limit on the permissible amount of negative curvature or anisotropy (either of which produces a more rapid expansion) if we are to have a universe with some structure. On the other hand, a universe with too much positive curvature would collapse before complex structures could have had time to develop. These considerations thus seem to isolate a relatively flat, isotropic universe from the class of nearly homogeneous universes, if we want to maximize the probability that "intelligent" structures develop.

The coincidence involving the age of our universe might then be "derived" from the restriction on the temperature. If we assume that the ratio of photons to baryons is fixed by the values of the fundamental constants of nature, then the temperature of *any* model universe satisfying the above restrictions on smoothness and flatness would fall to a value low enough for atoms to bind together only after a time a bit less than the age of *our* universe. In addition, when the age of the universe greatly exceeded this value, the expansion would have reduced the average temperature well below the optimal level for interactions among atoms (since stars would have consumed their nuclear fuel). Even though small pockets of higher temperature would exist, after a relatively short time they would cool below the temperatures required for significant chemical reactions to occur.

Arguments such as these have been cited by some as evidence for the validity of the Anthropic Principle. But even if we could make these arguments more precise, a fundamental difficulty would persist. Although in principle we might eventually claim to show by detailed calculations that complex structures could not arise in universes quite different from our own, we can never check this "prediction" because we can never observe even one of these other possible universes to see whether it contains intelligent life. The foundation of the scientific method—the ability to compare the predictions of a theory with the results of experiments and observations—is absent in any application of the Anthropic Principle. The uniqueness of the universe removes this principle from the realm of science and consigns it to the realm of speculation, unless we can find some other way to test its validity.

Nevertheless, it is difficult to ignore the Anthropic Principle: It hints at possible links between ourselves and the fundamental nature of the universe. Could the requirements for intelligent life be so strict that we could have known what the universe must be like without ever having observed the heavens? This follows if the Anthropic Principle is valid. By contrast, the laws of nature rather than the existence of life may eventually show that only our type of universe could exist. In this latter case, it might nevertheless seem remarkable that such a universe was able to produce structures capable of comprehending it.

Envoi

Humanity has made notable progress toward an understanding of the cosmos. We have overthrown many past misconceptions—often the result of earth-centered chauvinism—and are beginning to see how we fit into the vast and awesome universe that encompasses our past, present, and future. Although our present conception of the universe more closely corresponds to reality than did the beliefs of past eras, it almost certainly remains unimaginably naïve. Given the opportunity, the enormous potential of the human mind, activated by our innate desire to know, will continue to increase our scientific understanding.

As we extend the frontiers of knowledge, we deepen our awareness of the cosmos. This exploration—never ending yet deeply rewarding to the spirit—is one of the noblest of human endeavors. Future generations may judge us by our success in enlarging our horizons. As we strive to increase our knowledge, the key horizon lies within our minds.

READER'S GUIDE

THIS BOOK HAS SKETCHED the development of our view of the universe, along with those aspects of physics and astronomy that form the foundations of cosmology. We list below various references the reader should find useful in exploring more deeply some of the areas we have touched upon. Unless otherwise indicated, they are written at a level similar to that of this book.

Comprehensive surveys of cosmology are:

Harrison, Edward. *Cosmology: The Science of the Universe.* Cambridge: Cambridge University Press, 1981.
Silk, Joseph. *The Big Bang.* San Francisco: W.H. Freeman, 1980.

A less-detailed panorama is provided by:

Davies, Paul. *The Edge of Infinity: Where the Universe Came From and How It Will End.* New York: Simon and Schuster, 1982.

Greater emphasis on the very early history of the universe, conveying the excitement of recent discoveries in physics and astronomy, is provided in:

Weinberg, Steven. *The First Three Minutes.* New York: Basic Books, 1977.

More technical presentations of cosmology are:

Sciama, Dennis W. *Modern Cosmology.* Cambridge: Cambridge University Press, 1971.
Peebles, P.J.E. *Physical Cosmology.* Princeton: Princeton University Press, 1971.

Various epochs in the history of cosmology are treated in:

Munitz, Milton K., editor. *Theories of the Universe.* Glencoe, Illinois: The Free Press, 1957.
Dickson, F.P. *The Bowl of Night: The Physical Universe and Scientific Thought.* Cambridge, Massachusetts: M.I.T. Press, 1968.
Ferris, Timothy. *The Red Limit.* New York: William Morrow and Co., 1977.
Pannekoek, Anton. *A History of Astronomy.* New York: Interscience Publishers, 1961.

A survey of astronomy that explains clearly its physical basis is:

Shu, Frank H. *The Physical Universe.* Mill Valley, California: University Science Books, 1982.

At a somewhat less technical level than Shu's book is:

Goldsmith, Donald. *The Evolving Universe.* Menlo Park, California: Benjamin/Cummings, 1981.

An eclectic and provocative view of astronomy, with emphasis on the solar system and the development of life, is:

Sagan, Carl. *Cosmos*. New York: Random House, 1980.

A useful reference for all astronomical topics is:

Mitton, Simon, editor. *The Cambridge Encyclopaedia of Astronomy*. New York: Crown Publishers, 1977.

Spectacular photographs of our neighbors in space are found in:

Ferris, Timothy. *Galaxies*. San Francisco: Sierra Club Books, 1980.

An analysis of the way in which astronomical phenomena are discovered is presented in:

Harwit, Martin. *Cosmic Discovery*. New York: Basic Books, 1981.

An influential analysis of the roles of crises, paradigms, and revolutions in the progress of science is:

Kuhn, Thomas S. *The Structure of Scientific Revolutions*. Chicago: University of Chicago Press, 1970.

A thoroughly enjoyable and profound series of lectures on the fundamental nature of the physical world by one of the world's greatest physicists is found in:

Feynman, Richard. *The Character of Physical Law*. Cambridge: M.I.T. Press, 1965.

The following books explain a range of relevant topics in physics:

Davies, P.C.W. *The Forces of Nature*. Cambridge: Cambridge University Press, 1979.
Taylor, Edwin and John Wheeler. *Spacetime Physics*. San Francisco: W.H. Freeman & Company, 1966.
Geroch, Robert. *General Relativity from A to B*. Chicago: University of Chicago Press, 1978.
Davies, P.C.W. *The Search for Gravity Waves*. Cambridge: Cambridge University Press, 1980.

We list below (in chronological order within each category) some recent articles in *Scientific American* that may be useful to those interested in specific topics.

History

Gingerich, Owen. "Copernicus and Tycho." December 1973, p. 86.
Cohen, I. Bernard. "Newton's Discovery of Gravity." March 1981, p. 166.

Elementary-Particle Physics

Weinberg, Steven. "Unified Theories of Elementary-Particle Interaction." July 1974, p. 50.
Glashow, Sheldon Lee. "Quarks with Color and Flavor." October 1975, p. 38.
Wilson, Robert R. "The Next Generation of Particle Accelerators." January 1980, p. 42.
t'Hooft, Gerard. "Gauge Theories of the Forces Between Elementary Particles." June 1980, p. 104.
Wilczek, Frank. "The Cosmic Asymmetry Between Matter and Antimatter." December 1980, p. 82.

Georgi, Howard. "A Unified Theory of Elementary Particles and Forces." April 1981, p. 48.

Weinberg, Steven. "The Decay of the Proton." June 1981, p. 64.

Carrigan, Richard A., Jr. and W. Peter Trower. "Superheavy Magnetic Monopoles." April 1982, p. 106.

Gravitation

Thorne, Kip S. "The Search for Black Holes." December 1974, p. 32.

Hawking, S.W. "The Quantum Mechanics of Black Holes." January 1977, p. 34.

Chaffee, Frederic H., Jr. "The Discovery of a Gravitational Lens." November 1980, p. 70.

Weisberg, Joel M., Joseph H. Taylor, and Lee A. Fowler. "Gravitational Waves from an Orbiting Pulsar." October 1981, p. 74.

Astronomical Objects

Ostriker, Jeremiah. "The Nature of Pulsars." January 1971, p. 48.

Strom, Richard G., George K. Miley, and Jan Oort. "Giant Radio Galaxies." August 1975, p. 26.

Cameron, A.G.W. "The Origin and Evolution of the Solar System." September 1975, p. 32.

Kirshner, Robert P. "Supernovas in Other Galaxies." December 1976, p. 88.

Dickinson, Dale F. "Cosmic Masers." June 1978, p. 90.

Gorenstein, Paul, and Wallace Tucker. "Rich Clusters of Galaxies." November 1978, p. 110.

Strom, Stephen E. and Karen M. Strom. "The Evolution of Disk Galaxies." April 1979, p. 72.

Bok, Bart J. "The Milky Way Galaxy." March 1981, p. 92.

Blandford, Roger D., Mitchell C. Begelman, and Martin Rees. "Cosmic Jets." May 1982, p. 124.

Cosmology

Rees, Martin J. and Joseph Silk. "The Origin of Galaxies." June 1970, p. 26.

Schramm, David N. "The Age of the Elements." January 1974, p. 69.

Pasachoff, Jay M. and William A. Fowler. "Deuterium in the Universe." May 1974, p. 108.

Groth, Edward J., P. James E. Peebles, Michael Seldner, and Raymond M. Soneira. "The Clustering of Galaxies." November 1977, p. 76.

Muller, Richard A. "The Cosmic Background Radiation and the New Aether Drift." May 1978, p. 64.

Meier, David and Rashid A. Sunyaev. "Primeval Galaxies." November 1979, p. 130.

Barrow, John P. and Joseph Silk. "The Structure of the Early Universe." April 1980, p. 118.

Osmer, Patrick S. "Quasars as Probes of the Distant and Early Universe." February 1982, p. 126.

Gregory, Stephen A. and Laird A. Thompson. "Superclusters and Voids in the Distribution of Galaxies." March 1982, p. 106.

Observational Instruments

Giacconi, Riccardo. "The Einstein X-Ray Observatory." February 1980, p. 80.

Readhead, Anthony C.S. "Radio Astronomy by Very-Long-Baseline Interferometry." June 1982, p. 52.

Bahcall, John N. and Lyman Spitzer, Jr. "The Space Telescope." July 1982, p. 40.

ABOUT THE AUTHORS

Photo: Peter D. Miller

DONALD W. GOLDSMITH has written or edited eight books on physics and astronomy, including *From the Black Hole to the Infinite Universe* (with Donald Levy), *The Search for Life in the Universe* (with Tobias Owen), *Scientists Confront Velikovsky,* and *The Quest for Extraterrestrial Life: A Book of Readings.* He served as a consultant to the "Cosmos" series of television programs, which, he says, taught him more about Hollywood than he had planned to learn. During the past eight years, Dr. Goldsmith has taught astronomy at Chabot College, Chapman College, and the Irvine, Santa Cruz, and Berkeley campuses of the University of California.

Born in Washington, D.C., Dr. Goldsmith received an undergraduate degree in astronomy from Harvard College in 1963 and a Ph.D. in astronomy from the University of California, Berkeley, in 1969. After postdoctoral appointments at Stanford University and the University of California, Berkeley, he became an assistant professor at the State University of New York, Stony Brook, in 1972, a position from which he retired in 1974 to found Interstellar Media, which produces educational materials for scientific purposes.

Dr. Goldsmith lives in Berkeley and is now a visiting lecturer in the astronomy department of the University of California campus there. He says that his daughter, Rachel, now six, has considerably more curiosity about the universe than he did at her age—which he does not attribute to his books.

Photo: Leo Holub

Robert V. Wagoner's scientific direction was established rather suddenly in 1960, when he attended a series of lectures on cosmology by the British astrophysicist Sir Fred Hoyle. At the time, Wagoner was a mechanical engineering undergraduate at Cornell University with plans to pursue aeronautical engineering in graduate school. Born and raised in Teaneck, New Jersey, his other major interest was golf, leading (by hard work rather than native ability) to a position on the Cornell golf team.

Following Hoyle's lectures, Wagoner read every book on cosmology he could find. After graduation from Cornell in 1961, he entered Stanford University, where he recived an M.S. in engineering science in 1962 and a Ph.D. in physics in 1965. He then accepted a position as a research fellow at the California Institute of Technology. There he collaborated with William A. Fowler and Hoyle on the definitive calculation of the abundances of the elements produced in the first moments of the expansion of the universe. In 1968 Wagoner joined the faculty at Cornell as an assistant professor of astronomy. In 1973, he returned to Stanford, where he is professor of physics.

Professor Wagoner is internationally recognized for his theoretical work in gravitation (particularly gravitational radiation) as well as in many areas of astrophysics and cosmology. At present he is investigating new ways to learn about the early universe and to measure the expansion rate of the present universe. He feels a great debt to Leonard Schiff (his thesis advisor), William Fowler, and Edwin Salpeter (with whom he worked at Cornell) for their influence on his career.

His academic honors include awards of a Sloan Foundation Fellowship and a Guggenheim Foundation Fellowship. In 1976 he was a Sherman Fairchild Distinguished Scholar at the California Institute of Technology, and in 1978 he was the George Ellery Hale Distinguished Visiting Professor at the Enrico Fermi Institute of the University of Chicago. Wagoner was also a Visiting Fellow at the Institute of Theoretical Astronomy in Cambridge, England, during the summers of 1967 and 1971.

Professor Wagoner lives on the Stanford campus with his wife, Lynne, and his daughters, Alexa and Shannon. His athletic interest has evolved from golf through tennis to running.

CREDITS

COVER   The cover photograph depicts the central region of the Virgo cluster of galaxies, about 50 million light-years from us. Photograph courtesy of Kitt Peak National Observatory, Cerro Tololo Inter-American Observatory. Cover design by Sharon H. Smith.

FIGURES   All drawings in this book are by Donna Salmon, Vallejo, California.

FRONTISPIECE   The shorter exposure photograph (top photo) shows the double quasar Q0957 + 56. The two quasar images, which are only six seconds of arc apart, have the same redshift (1.41) and identical spectra. The longer exposure (bottom photo), which shows precisely the same region of sky to a fainter level, is dominated by a rich cluster of galaxies at a redshift of 0.36. In the deeper exposure, the lens galaxy can be seen as a fuzzy, elliptical image that completely blots out the lower image of the quasar and part of the upper image as well. From analysis of these pictures and other data, it has been found that the double quasar is really two images of a single quasar, caused by multiple focusing of its light as it shines through the intervening galaxy. The light from the quasar is bent (in two directions) by the gravitational field of the intervening galaxy, just as starlight has been observed to be bent by the gravitational field of the sun (see p. 106).

These pictures, in which the lens galaxy was first seen, were taken by J. Kristian, J.A. Westphal, and P.J. Young at the 200-inch telescope on Palomar Mountain, using a CCD (charge-coupled device) solid state television imager developed by Texas Instruments. A similar camera will be used on the Space Telescope, which is due to be carried into orbit around the earth by the Space Shuttle in 1985. The lower exposure is one of the deepest astronomical photographs ever taken. (Palomar Observatory photograph, courtesy of Dr. Jerome Kristian, Mt. Wilson and Las Campanas Observatories.)

INDEX

Universe:
   age of, 27, 104, 105, 126–28, 133,
      134, 141, 168, 172, 180–81, 183
   causal, 159–60
   chaotic, 169
   closed, 100–1, 121, 140
   curvature of, 100–2, 104, 114–26,
      140, 146, 155, 168, 181–83
   deceleration of, 123–26, 167–68
   definition of, 1
   early, 137–61, 169
   expansion of, 27, 32, 52, 77, 103–5,
      117–23, 125–28, 133, 134, 138,
      139, 147–50, 155, 159, 167, 168,
      173
   geometry of, 100–2, 113–14
   homogeneity of, 31, 98–99, 113,
      117, 131, 166, 183
   initial conditions of, 169, 180
   isotropy of, 96, 99, 113, 166, 183
   open, 100, 101, 121, 155

   temperature of, 129–30, 134–35,
      138
Virgo cluster, 63
Virtual particle, 38
VLA (Very Large Array), 178, 179
Wavelength, 49–52, 107–9, 119, 123,
   170–79
Weak interaction, 38, 39, 43, 143, 144
Weinberg, Steven, 44
Whirlpool Nebula, 29
Wilkinson, David, 110
Wilson, Robert, 106–8
Wirtz, Carl, 103
Woody, David, 108
World line, 156–60
W boson, 36, 38, 39, 44, 143, 144
X boson, 144–46
X-ray astronomy, 7, 8, 32, 69, 90–91,
   96, 99, 135, 172–74
Z boson, 36, 38, 39, 44, 143, 144
Zweig, George, 37